ROBOTS WON'T
SAVE JAPAN

A volume in the series

The Culture and Politics of Health Care Work

Edited by Suzanne Gordon and Sioban Nelson

For a list of books in the series, visit our website at cornellpress.cornell.edu.

ROBOTS WON'T SAVE JAPAN

An Ethnography of Eldercare Automation

James Wright

ILR PRESS

AN IMPRINT OF CORNELL UNIVERSITY PRESS   ITHACA AND LONDON

Copyright © 2023 by Cornell University

All rights reserved. Except for brief quotations in a review, this book, or parts thereof, must not be reproduced in any form without permission in writing from the publisher. For information, address Cornell University Press, Sage House, 512 East State Street, Ithaca, New York 14850. Visit our website at cornellpress.cornell.edu.

First published 2023 by Cornell University Press

Librarians: A CIP catalog record for this book is available from the Library of Congress.

ISBN 978-1-5017-6804-0 (hardcover)
ISBN 978-1-5017-6806-4 (pdf)
ISBN 978-1-5017-6805-7 (epub)

To my parents

Contents

Acknowledgments	ix
Note on Transliteration	xi
Introduction	1
1. Crisis and Care Robots	21
2. Developing Robots and Designing Algorithmic Care	36
3. Portrait of a Care Home	58
4. Hug: Reconfiguring Lifting	80
5. Paro: Reconfiguring Communication	95
6. Pepper: Reconfiguring Recreation	115
7. Beyond Care Robots	133
Notes	151
References	163
Index	175

Acknowledgments

I would like to thank the robotics engineers and staff at the Robot Innovation Research Center, part of the National Institute of Advanced Industrial Science and Technology (AIST), for their kind cooperation and assistance with my fieldwork. In particular, I would like to thank Matsumoto Yoshio for his warmth and generosity in extending me an invitation to conduct research at AIST and for spending a great deal of time explaining how the institute worked as well as the research projects under way there. Despite differing perspectives on the particular project of developing care robots, I have no doubt of the sincere desire of all of the researchers I spoke with at AIST to improve people's lives with technology. I would also like to thank Mr. K and the care staff and residents at Sakura, who were incredibly welcoming and patient in putting up with me hanging around and asking questions—particularly given the often-severe constraints on their time. Mr. K was candid, open, and willing to explain his thinking and indulge my questions on all aspects of life and work in a care home, and I am extremely grateful for all his help. My heartfelt thanks also go to all the other people I spoke to and interviewed for this project, including staff at the Ministry of Health, Labour, and Welfare, the Japan Agency for Medical Research and Development, the Japan Quality Assurance Organization, the Association for Technical Aids, Fuji Machine Manufacturing, SoftBank, Atelier Akihabara, PIP&WiZ, Orix Living, Toyota, and KDDI au.

Thanks to everyone at the Hong Kong Institute for the Humanities and Social Sciences at the University of Hong Kong who provided invaluable feedback on my PhD thesis from which this book is adapted. In particular, I am immensely indebted to my supervisors Gonçalo Santos and Nakayama Izumi, who helped guide and support my project at every turn. I would also like to thank the members of my examination committee, Jennifer Robertson, Cheris Chan, and Nakano Yoshiko, who provided a great range of constructive feedback and advice. The PhD research that underpins this book was supported by the Hong Kong Postgraduate Fellowship, and my fieldwork in Japan was undertaken with the financial assistance of the Konosuke Matsushita Memorial Foundation and the Sin Wai-Kin Junior Fellowship. I am also very grateful to Glenda Roberts for kindly hosting me at Waseda University's Graduate School of Asia-Pacific Studies during my fieldwork.

My ongoing research benefited greatly from time spent at the Fondation France-Japon de l'EHESS, and I am deeply grateful to Sébastien Lechevalier and his colleagues for helping make my stay such a beneficial experience. Sue Yeandle from the University of Sheffield also very kindly read my manuscript and provided helpful advice and encouragement.

I would like to thank Sioban Nelson and Suzanne Gordon, coeditors of the ILR Press series *The Culture and Politics of Health Care Work* to which this book is a contribution, as well as Ellen Labbate, Ange Romeo-Hall, and Jim Lance from Cornell University Press, and Anne Jones from Westchester Publishing Services for all their help. I am also very grateful to the anonymous reader who reviewed my manuscript and provided such constructive and helpful feedback. Huge thanks to Russell Henshaw and Ryszard Auksztulewicz for their invaluable advice and detailed comments on chapter drafts, as well as to Odagawa Daiki, Komamiya Shunsuke, and Kusanagi Kanako who have very kindly and patiently answered my endless stream of Japan-related questions over the years.

Finally, a special thanks to my family who have supported me throughout this journey.

Note on Transliteration

All translations are my own. Japanese is transcribed using the modified Hepburn system, and Japanese names are written according to Japanese name order (surname first), unless the person's name is well known in the reverse order.

A rough exchange rate of ¥100 to US$1 is provided throughout.

ROBOTS WON'T SAVE JAPAN

Introduction

> A Japanese care home at night. An older man wanders along an empty corridor, looking confused, until he comes face to face with "Pepper," a four-foot-tall white plastic-and-metal humanoid robot, with big, illuminated eyes and a high-pitched chirpy voice. As the man walks up to the robot, it starts talking to him: "Good evening! Where are you off to so late?" and asks him a series of questions: "When's your birthday? Tell me something you can be proud of. Where's your home? Who do you live with?" Meanwhile, the robot covertly takes a photo of his face and emails it to a care worker, who rushes to the scene and escorts the man back to bed.

(Description of a staged 2016 promotional video for a "monitoring" app for the robot Pepper[1])

> A Japanese care home in daytime. A care worker, Fujita, spends several minutes booting up the same type of humanoid robot Pepper in the room where it is stored, and when it is ready to go, she wheels it into the corridor. She helps an older female resident, Suzuki, toward the robot, and quickly carries a chair over for her to sit on so that she is positioned in front of it. Suzuki seems excited to interact with Pepper. Fujita uses the robot's touch screen to select its quiz app, and it begins the quiz with Suzuki. At first, everything seems to go well, and Fujita goes off to do other tasks, although she continues to keep an eye on Suzuki's interaction with the robot. But the background noise of the care home begins to interfere with Pepper's speech recognition. It asks Suzuki, "What's four times six?" She answers: "Twenty-four." There is a pause. She repeats, "Twenty-four." Pause. "Twenty-four!" "Wrong answer." "Twenty-four!!" "Wrong answer." "What do you mean it's the wrong answer, it's twenty-four!" "Wrong answer." Eventually, Suzuki loses patience with the robot and starts to get up. Fujita runs over to help her up and take her back to her usual seat, and then immediately wheels Pepper back to its storage area in case another resident trips over it. She says to me with a wry smile, "How much does this thing cost again?"

(Pepper in use at a nursing care home in Japan in April 2017, description based on field notes)

This is a book about robots, Japan, and the future of care. It is about state and corporate attempts to bring together increasingly sophisticated, but also exceedingly hyped, robotic technologies with the unprecedented number of older people requiring care. It is also a book about what it means to do care work with robots. Are robots an ideal solution to the "problem" of aging, or do they create new problems of their own? Could a machine, or a suite of machines, partially or fully replace the need for human care workers? If so, what would become of care and care work?

Although this book focuses on Japan, the questions it asks and hopes to answer are critical for many other higher income countries that are contending with growing populations of older people but do not seem to have enough caregivers to look after them. Japan may appear to be a special case since it is facing such a serious care crisis. The statistics seem to paint an extraordinary picture. By 2050, the proportion of people over sixty-five years of age is forecast to grow to nearly 40 percent of the overall population, with 514,000 people over one hundred years of age—nearly *ten times* the number in 2016. In 2000, there were about four adults of working age for every person over sixty-five; by 2050, there will be near parity. By the same year, almost one in ten of the population—nine million people—is expected to be living with dementia. To cope with the level of care that will be needed, Japan's government expects that by 2040, one in ten adults of working age will be employed as a care worker, up from around one in 160 in 2000, not counting the many people providing informal care to older family members and friends. In 2012, annual social benefit spending for older adults was at an all-time high—¥74.1 trillion ($741 billion), representing 21 percent of national income. By 2040, it is forecast to hit ¥190 trillion ($1.9 trillion).[2]

Japan is, however, not alone in this predicament. In Europe, it is estimated that one-third of the population will be over sixty-five years of age by 2060, leading to a doubling of care needs for older adults compared with the level in 2012. In the United States, the figure is expected to be slightly lower, with around one-quarter of the population over sixty-five by 2060, although a sense of crisis is even more immediate due to the absence of state assistance for eldercare.[3] Many European Union countries as well as the United Kingdom expect significant and growing shortages of care staff in the near future. This is why policymakers and nursing home administrators across Europe and North America are looking to Japan for possible technological solutions to their caring crisis.

In attempts to preempt some of the vast social and economic transformations that these predictions seem to imply, some in Japan and other aging countries have been imagining alternative futures. Over the past twenty years, one such vision has increasingly materialized among government technocrats, research managers, and corporations. It is a future populated by robots that care, and it has driven Japan to pioneer a new kind of techno–welfare state increasingly focused on high-tech systems and networked digital devices. In this technology-centered reimagining of care, machines with names like Hug, HAL, and Pepper are intended to supplement or replace the work of human caregivers while also assisting older people to look after themselves, breaking down the tasks of care into linear strings of simple, repeatable physical actions and speech that can be digitally reproduced by computational algorithms and performed by robots.

Japan is an ideal case study in helping us explore the fantasy of robots solving the significant social and demographic challenges that confront aging postindustrial societies. The idea of robots coming to the rescue taps into a rich invented tradition of Japan as "robot kingdom" which has been cultivated over decades, with relentless promotion of robots in popular culture and across state, media, and industry. Yet despite considerable domestic and international interest in eye-catching Japanese robots, actual knowledge about what kinds of technologies are being developed, how, and for what purposes remains scant. Many news stories portray an exoticized version of Japan that fails to capture the complex reality of how care robots are developed and used, while many academic studies of robots are conducted by those with significant stakes in the success of the industry.

This book explores the world's single largest project to date aimed at developing and implementing care robots, launched by the Japanese government, as a lens through which to view this attempt to transform the future of care. It examines the differences between how the roboticization of care is imagined, how it is proceeding, and what this process looks like from the perspective of an eldercare facility and care workers expected to use robots. In doing so, it grapples with a number of questions: What do robots mean for the future of care? Will robots decimate human employment in the care sector, and would this in fact be a desirable outcome, given the often-negative reputation of caring jobs? What is the interface between the engineers and technocrats who design and promote robots, the workers who use them, and the care home residents whose lives they are ostensibly designed to improve? How do robots contribute to transformations of care labor and practices, what it means to care, and how those doing this work understand good care?

I will argue that despite considerable hype, lofty expectations, and substantial investment, robots alone cannot yet deliver on the promise of solving care crises in Japan or elsewhere. Efforts to develop and implement them instead call attention to what might be lost in the roboticization of care, while also revealing alternative approaches that go beyond technological fixes.

Caring for Capitalism

I first became interested in Japanese robots in 2007 during my master's degree at Oxford University. The Japanese government had just released *Innovation 25*, an optimistic vision statement that imagined everyday life in Japan in the year 2025—a society full of robots and other high-tech gadgetry. An ambitious new research project aiming to develop a robot version of a human child (CB2) had also been launched, and other Japanese robots were garnering international media exposure on a regular basis. It seemed as if the futuristic "robot society" depicted by the Japanese government was just around the corner. My enthusiasm was not shared by my adviser, an anthropology professor, who counselled me not to be distracted by the hype and fanfare of robots that were rarely used in daily life. But by the time I began my PhD in 2014, academic interest in robots was continuing to build. The relatively new field of human-robot interaction studies was expanding rapidly, and scholars across the social sciences showed more interest than ever in the subject. Japan was a key center for the development of interactive and humanoid robots, and the government had just initiated the largest ever care robot development and implementation project. If care robots had not entered everyday life quite yet, they seemed poised to do so.

In Japan, state and industry backers of care robots imagine them as helpful companions, looking after elderly parents and even young children at home, and enabling working-aged members of their family, particularly women, to keep their jobs and pursue economically productive lives without having to spend time and effort providing care. These technocrats aim to promote the development and use of a range of what they call personal care robots, to be employed mainly in care institutions but also increasingly in private homes, to help with demanding or time-consuming care tasks. Some of these devices are aimed at physical care, including machines that can help lift older people unable to get up by themselves, assist with mobility and exercise, monitor their physical activity and detect falls, feed them, and help them take a bath or use the toilet. Others are aimed at engaging older people socially and emotionally in order to manage, reduce, and even prevent cognitive decline, provide companionship and therapy for lonely older people and their hard-pressed caregivers, make those with challenging behavior due to dementia-related conditions easier for care staff to manage, and reduce the number of caregivers required for day-to-day care.

In promoting high-tech robotic solutions to the seemingly technical problems of care, such as the transfer of older people from bed to wheelchair which contributes to widespread back pain among Japanese caregivers, government and industry supporters also hope to improve poor employee hiring and retention rates. In this way, they expect to be able to capitalize on peripheral elements of the labor force by encouraging older or less physically able workers into the formal care sector. At the same time, they aim to support informal familial care at home and aid the "independence" of the increasing proportion of older people who live alone, thereby mitigating the ever-growing demand for institutionalization.[4] Some believe that introducing care robots to do many of the tasks currently performed by human caregivers can reduce the burden of care on families, alleviate labor shortages, reduce the need for politically problematic migrant care labor from other parts of Asia, and increase productivity in the sector, thereby cutting spiraling care costs and revitalizing the economy. All this while reducing loneliness, preventing elder abuse, and improving standards of well-being and care for older people who require it—although these latter aims tend to be presented as, at best, an afterthought. Care robots are intended to showcase Japanese innovation and technology and to create a massive new high-tech export industry for Japan, supplying an international market that will continue to grow rapidly for decades in tandem with the aging populations of wealthy countries. The imagined benefits of care robots are seemingly endless.

The reason for the care crisis in Japan seems straightforward: there are too many older people who need care and not enough younger people to provide it. Yet, this simple statement obscures layers of political, economic, and sociocultural complexity encompassing ideologies and politics of neoliberalism and globalization, gender and family relations, and intergenerational ethics. If this is a crisis of care at a particular historical moment of demographic extremes, it also seems to be a broader crisis of capitalism. Capitalism requires workers; workers have to be reproduced—not only biologically but also socially: looked after, cared for, brought up, and socialized through countless interactions with family, friends, and other caregivers throughout their lives, most of which is done informally and in the home—unpaid, out of love, duty, or other reasons unrelated to money (Folbre 2001). Without this reproductive labor, capitalism could not exist: as Joan Tronto notes, "No state can function without citizens who are produced and reproduced through care" (2013, 26). Care work, including eldercare work, can be understood as reproductive labor in the sense that care reproduces society and social values, maintaining social cohesion and intergenerational reciprocity. Reproductive labor historically has been undervalued or not valued at all in economic terms; care labor has until relatively recently been taken for granted.

Nancy Fraser has argued that the crises of care seen across many postindustrial societies today originate from this "contradiction" of capitalism that "tends to destabilize the very processes of social reproduction on which it relies" (2016, 100), while Anna Tsing (2015) argues that the translation of noncapitalist areas of value (including, I would suggest, unpaid caring) into capitalist systems is in fact an essential aspect of how capitalism works, which, like Fraser, she sees as unsustainable in its current form. This process only seems contradictory because of the rhetoric of rationalization used to describe markets and economies, which tends to ignore noncapitalist elements such as care that are among its vital inputs. Crises of care occur as the tacit economic need for unpaid care labor conflicts with the competing drive for maximal employment in the formal labor force. Capitalist models that largely ignore care labor therefore start to look increasingly precarious when reproduction rates fall and large sections of the population pass retirement age.

Whereas most Western countries have turned to the near-term solution of immigration from lower income countries to provide a fresh supply of low-paid care workers, the situation in Japan has been complicated—uniquely among higher income countries—by relatively low levels of immigration, for sociocultural and political reasons that will be examined later. Nevertheless, amid growing international recognition that "free" familial care is actually paid for by the individual and societal economic opportunity costs of lower household income and fewer employees in the workforce, the Japanese government has been pushing eldercare from mainly informal and unpaid provision into the paid sector in hopes of freeing up family carers to participate in the formal economy. In the process, it has exposed the substantial economic costs of trying to bring informal unpaid care labor into the formal capitalist economic system through public financing on a national scale. Yet, this formalization of labor is also creating new markets. The world of care, long hidden in plain sight, has now become a crucial area of commodification and "value creation"—a new market frontier ripe for innovative disruption, with older people reconceptualized as a source of economic growth and rejuvenation rather than a burden on national resources. Sociologist Ito Peng (2018) calls this the care economy, arguing that care has become "a key driver of the new service sector economic growth and expansion" across almost all high- and middle-income countries.[5]

The development of care robots represents an attempt to square the apparent contradiction of capitalism in Japan while continuing to grow the economy: to digitally and mechanically reproduce and replace large swathes of human care, substituting capital for labor and so acting as a kind of bridge between productive and reproductive labor. Use of robots seems the logical extension of efforts to rationalize human work across all sectors and particularly those such as care work that are highly labor intensive by engineering a replacement that is entirely commodified—creating what robotics scholars Noel Sharkey and Amanda Sharkey (2012) have described as akin to an eldercare factory.

The aim of developing care robots is to make eldercare sustainable within the context of Japan's export-driven economy—providing more efficient care services domestically while at the same time globalizing robot care by standardizing and exporting robotic technologies and new scalable and interchangeable robotic practices of care to other aging societies around the world. In combining older industrial robotic technologies (including hardware components such as robot arms, actuators, conveyor belts, and robotic vision systems) with newer cutting-edge developments (particularly those pioneered and developed through smartphones—including apps, sensors, smaller and faster processors—as well as the networked internet of things, ubiquitous data and telecommunication infrastructures, and artificial intelligence (AI) techniques such as machine learning), Japanese engineers aim to open up a new market for existing industrial robot and robot component makers, as well as high-end materials manufacturers and software engineers. A focus on robots would seem to play to the strengths of Japan's existing research environment and engineering and manufacturing capabilities and could reassert and showcase its dominance in the wider global field of service robotics, which is increasingly being threatened or overtaken by aggressive, innovative, and state-backed competition from regional rivals like China and South Korea. The title of a 2006 book by Nakayama Shin, at the time board chairperson of Yaskawa Electric Corporation, a leading industrial robot manufacturer that also makes medical and nursing care robots, summed up this belief in a robotic panacea: *Robots Will Save Japan*.[6]

The Necessity of Robots and Nostalgia for Modernity

Robots are the ultimate enchanting dream of modernity, embodying progress through innovation. No wonder commentators in the United States and Europe have occasionally adopted apocalyptic language to describe how processes of roboticization and automation are bound ultimately to end human work—like a techno-utopian version of Francis Fukuyama's infamous prediction of the end of history.

For many, however, this vision is closer to a techno-dystopia. Western media reports about robots have largely been dominated by anxieties about mass job losses. Fears in North America and Europe of roboticization and automation leading to mass unemployment reached widespread public consciousness in 2013 with the release of a report by the Oxford Martin School that suggested half of all jobs in the United States at the time were at risk of being lost to computerization (Frey and Osborne 2013). The publication of several popular books and reports on the subject followed in quick succession, including Erik Brynjolfsson's *The Second Machine Age* (2014) and Martin Ford's *Rise of the Robots* (2015). A 2017 report by the McKinsey Global Institute predicted up to eight hundred million workers globally—including one-third of American workers—would lose their jobs by 2030 due to automation (Manyika et al. 2017). In the late 2010s, however, the imminent destruction of jobs was not forthcoming, and fears about large-scale unemployment due to roboticization and AI seemed to have receded, at least temporarily. Historically low unemployment rates in the United States in early 2020 were followed months later by historic highs due not to robots or AI but to the sweep of COVID-19. The pandemic underlined the crucial importance and current irreplaceability of many low-paid person-to-person jobs in health and nursing care, supermarkets, pharmacies, logistics, public transport, garbage collection, and delivery services.[7] Nevertheless, tech utopians and tech dystopians alike continue to share a narrative that roboticization and AI will ultimately destroy repetitive and manual jobs, although optimists believe these will be replaced by higher value "creative" jobs better suited to human potential or that the enormous profits generated by the owners of these technological means of production will be redistributed via some kind of universal basic income to the unemployed majority.

If roboticization has been looked on with dread by many in the United States and Europe as a seemingly unstoppable process that appears destined to undermine economic and social stability and lead to precarious and unpredictable futures, in Japan a different kind of transformation is envisaged. Speculation about potential negative consequences of the introduction of robots was largely absent from Japanese media reports in the 2000s and 2010s, which tended instead to celebrate the achievements of robotics engineers and showcase promising examples of the trialing and implementation of robots in a wide variety of settings—invariably framed within the context of the aging population and shrinking workforce. Japanese worries tend to center less on job losses than on labor shortages. Nevertheless, many visionaries in Japan share with their more skeptical Euro-American counterparts a belief in the inevitability of robots. In fact, the argument in favor of robotic solutions to Japan's problems has proved so powerful that robots are often discussed with the language of necessity.

In an interview with Bloomberg in 2015, Yurugi Yoshiko, a robotics expert at Japan's national research and development agency, the New Energy and Industrial Technology Development Organization (NEDO), asserted that "we are entering an era when we will definitely have to rely on the help of robots" (Bremner 2015). Similarly, a report prepared for the Japanese government in 2018 notes that "Japan . . . has a high cultural affinity for humanoid robots. . . . Owing to the growing issue of declining birthrates, it is a technology Japan requires. . . . Nursing robots and autonomous vehicles are expected to be direct solutions to an aging society" (AIR 2018, 24–30). This language is also commonly adopted in cross-disciplinary international scholarship on care robots. For example, Florian Kohlbacher and

Benjamin Rabe argue that "the logic of demographic ageing seems to create a natural demand for such technologies [i.e., care robotics]" (2015, 22), while anthropologist Michael Berthin states that "since Japanese both need and like robots, it is inevitable that they will soon be a major part of Japanese society" (2014, 15). One Japanese robotics engineer I interviewed told me, "Care robots . . . I think they're very important. If we don't use these kinds of convenient things, the world of care won't be able to manage. So I think they'll definitely become necessary." A second told me, "It's a difficult question, but I think as a robot developer, we just have to do it, because the shortage of carers will get gradually worse as more people need care." Another described care robotics as "a necessary technology." For many, robot care appears to have assumed an air of natural inevitability that forecloses other possibilities. Yet, robots are not a straightforward "natural" solution autogenerated by the market or robotics engineers but the product of sustained Japanese state support for and guidance of the robotics industry over decades.

The idea of using robots for eldercare first emerged in Japan in the mid-1990s, largely as a possible use case for interactive "next-generation" robots that were starting to be developed out of a context of previous research into industrial robots, robots for hazardous environments, prosthetics, humanoids, and toy robots (Wagner 2013).[8] Indeed, many of the robots used for eldercare were not originally designed or intended for this application and were repurposed from other roles.[9] An influential 2001 joint Japan Robot Association and Japan Machinery Federation report argued for the need to expand Japan's competitiveness in robotics beyond manufacturing, estimating that by 2025 the global market for medical and welfare robots would reach ¥1.1 trillion ($11 billion) (JARA/JMF 2001), and this optimistic estimate seems to have kickstarted significant and sustained government investment. In 2014, Japan's Ministry of Economy, Trade, and Industry (METI) estimated that the domestic market for nursing care robots alone would expand to ¥391 billion ($3.91 billion) by 2025 (Nikkei Asia 2014). Robots that care for older adults have become a central component of the Japanese government's broader robotics strategy and an area in which the government and industry have invested hundreds of millions of dollars over the past twenty years via a series of large-scale development and dissemination projects outlined in chapter 1. Although eldercare may initially have been an accidental or secondary application of robots, substantial public funding was now being directed at developing more and scaling up their implementation.

In 2015, Prime Minister Shinzō Abe announced the establishment of the Robot Revolution Initiative, promising to "spread the use of robotics from large-scale factories to every corner of our economy and society" (Bremner 2015) and branding 2015 "year one of moving towards a robot society" (DeWit 2015). This was to be a revolution that would not destroy human employment or undermine social cohesion but, on the contrary, lead to a more efficient robot society that would secure Japan's future on the world stage while paying for its mounting welfare bills. In the same year, the Headquarters for Japan's Economic Revitalization, headed by Abe, released a National Robotics Strategy for Japan which, among other goals, aimed to expand the market for nursing care robots to ¥50 billion ($500 million) by 2020 while eliminating widespread back pain caused by care work. The government's 2016–21 strategic plan for science and technology continued to emphasize an expansion of the development of robotics, information and communications technology, and AI, as well as their vital role in building what it describes as Society 5.0,[10] particularly in the field of nursing care (Cabinet Office 2019). Care robotics had become a central pillar of Japan's national strategy for economic "revitalization."

It may seem strange that a conservative political party that has been in power more or less continuously for the past seventy years should be promising revolution. Yet, Abe was a key proponent of a robot-centric economic strategy, which he shaped to suit his technocratic but conservative political platform. It hearkened back to the golden age of Japanese postwar economic growth through industry, while looking to innovative technological sources of national income and pride—a mixture of past and future visions of Japan that anthropologist Jennifer Robertson aptly sums up with the phrase "'retro-tech,' or advanced technology in the service of traditionalism" (2010, 28). In her 2018 book *Robo Sapiens Japanicus*, Robertson demonstrated how Japanese humanoid robots, despite their sophisticated technology and futuristic, progressive image, were used by Abe and his administration to embody and reinforce traditionalist, ethnonationalist and ableist views on gender, the family, race, and disability. She argues that these humanoid robots, imagined by some politicians to bolster their ideals of the traditional family, represented a "future past": renovation rather than innovation, or what we might call a nostalgia for modernity.[11] Care robots, too, are imagined to be technologies of nation rebuilding and renewal.

Although robots were favored by Abe since his first term as prime minister from 2006–7, Japan's care robot strategy has been masterminded by METI, financed by national funding agencies NEDO and the Japan Agency for Medical Research and Development (AMED), and operationalized by METI's research arm, the National Institute of Advanced Industrial Science and Technology (AIST). METI intended robotic products developed in Japan to be funneled through the huge care goods and services market created by the introduction of the national Long-Term Care Insurance system in 2000 and to be exported to welfare states and eldercare markets around the world. The global economic ambitions of this national robot care strategy were signaled by the fact that projects were initiated and led by METI, rather than by the Ministry of Health,

Labour, and Welfare, which might be expected for a project aimed at improving care, or even the Ministry of Education, Culture, Sports, Science, and Technology, which in theory is in charge of technology policy.

The largest care robot research and development project to date under Abe's state strategy was the ¥12.5 billion ($125 million) Project for the Development and Promotion of the Introduction of Robot Care Devices (*robottokaigokikikaihatsu, dōnyūsokushinjigyō*; hereafter Robot Care Project), run by AIST under the guidance of METI, NEDO, and AMED from 2013 to 2017. This hugely ambitious project to support the development and dissemination of robots to carry out tasks across almost every aspect of eldercare, including robots specifically developed for this work as well as those repurposed for it, had just started when I began to explore care robotics and became the main focus of my PhD research. Projects at the local and national levels continued after this one ended—for example, through a follow-on project titled the Project for the Development and Standardization of Robot Care Devices, which was due to end in 2021.

Despite the importance of official strategies and projects, it is equally important to note that not everyone in Japan, including in the government or even among engineers directly involved in these projects, shared a robotic vision for the future of care, as will become clear through the course of this book. Moreover, the reality of actually developing and using care robots is far more complex than straightforward utopian visions of the future might suggest. Robots are not stand-alone devices but rather component parts of interconnected sociotechnical assemblages that extend across diverse global actors. These systems are adopted in unexpected ways with unforeseen results when repurposed from other use contexts or when removed from the labs where they are created and tested, and then released "into the wild."

Disentangling Robots

What is a robot? The term derives from the Czech word *robota*, meaning "forced serf labor,"[12] and was coined by Karel Čapek and his brother Josef in Karel's 1920 play *R.U.R.* (standing for *Rossum's Universal Robots*) in which robots are synthetic versions of humans made out of a protoplasmic paste and are put to work in factories and offices. In its broadest sense, the category of robot seems to include almost any physical, virtual, or imagined machine (or metaphorically, software program or person) that can interact with its environment in some way. A more exact definition is difficult to pin down. Many technologies described as robots seem to have no shared form, characteristics, or purpose, with the term linking together a wide array of devices with varying functions, appearances, and components, used in a variety of industries and research projects, with or without some form of AI, and involving varying degrees of autonomy. When I asked several robot engineers how they defined robots, I received a different answer from each of them.

The difficulty of defining robots is partly due to the constantly changing character of the field of robotics. But robots themselves can sometimes stop being robots. During one of my first visits to AIST, I put the question of definition to Matsumoto Yoshio, the head of the Robot Innovation Research Center. He indicated the difficulty of answering by explaining how the printer scanner in his office, which few people would describe as a robot, meets many of the criteria that are commonly used to define one. After all, it is an aggregation of motors, sensors, and processors and operates semiautonomously—for example, by automatically sending data to its manufacturer signaling when its sensors determine that it should be serviced. His point was that robots often cease to be called robots when they leave the lab but instead become end products: vacuum cleaners, drones, smartphones. Robots often seem to have a temporal identity associated with open-endedness before their identity is stabilized in the form of a particular "finished" product. If, as often happens, a fully commercialized end product is not eventually produced, its identity remains unstable—a prototype continuously on trial. As anthropologist Christina Leeson described in her study of how the humanoid robot Telenoid was brought over from Japan and adapted for use in Danish care institutions, "the robot was less an object than a process; it was a becoming—not an entity.... The robot was constantly and reciprocally in the making as it transcended its laboratory to enter the daily lives of people around the world" (2017, 7).

Another defining characteristic of robots is their assembled nature. Robots are usually complex technological assemblages. A senior engineer at AIST told me that robotics is "like composite art, gathering a collection of possible engineering techniques to make things." In Japan, the field is itself a composite—robotics is not taught at college as a distinct major but exists as a synthesis of mechanical and electrical engineering, computer science, and increasingly also of psychology, cognitive and behavioral science, and the social sciences in the form of human-robot interaction studies. Robots and robotics incorporate a contingent, open-ended, ambiguous, diverse, and complex portfolio of established and emerging technologies in varying and often overlapping configurations, assembled from components such as sensors, processors, actuators, and manipulators, which sense and dynamically interact with their physical environment: acting *in* human times and spaces and *like* humans or other living creatures.[13] This often includes the performance of physical actions or speech analogous to labor performed by humans, as in the case of industrial robots used in factories, or care robots.

This does not tell the whole story, however. The assembled nature of robots goes beyond their mechanical and digital components. The term *robot* conjures up a plethora of colorful and fragmented images and ideas: for Euro-American audiences, the enduringly popular *Terminator* movies, *Westworld*, *Black Mirror*, *Real Humans*, Hanson Robotics' Sophia, industrial robot arms, Roomba vacuum cleaners, sex robots, and so on; for many in Japan, the list also includes manga and anime series such as *Astro Boy, Mazinger Z, Doraemon, Neon Genesis Evangelion,* and *Gundam*, Honda's ASIMO humanoid, as well as roboticist Ishiguro Hiroshi's famous lifelike androids such as Matsukoroid. Again, robots seem to be characterized by an enchanting quality of semiotic open-endedness—a resistance to a singular definition, and a capacity to carry heterogeneous and even contradictory cultural meanings. For this reason, robots are sometimes described as "boundary objects" or "multistable artifacts" in the language of science and technology studies, taking on different meanings for different people and in different cultural contexts.

This combination of technological and interpretive flexibility enables competing visions of the future to be projected onto robots. Robotics engineers sometimes complain about the negative connotations (particularly in the United States and Europe) of the term *robot* and about the unrealistic expectations this cultural baggage generates among the general public. The complex assemblages of sensors, motors, software, and other components termed "robots" that they work on are inextricably entangled with the vagaries of pop cultural depictions of sci-fi robots and media whims and (mis)representations. Their funding is often dependent to some extent on the tenor of such representations. Yet, engineers also actively and often enthusiastically exploit these pop culture associations to maintain and build hype to secure funding and media interest for their projects and grow brand awareness among the public, as well as to make sense of or give meaning to their own work. For example, it is no coincidence that a company in Japan called Cyberdyne (the name of the corporation that produces the killer cyborgs of the *Terminator* series) produces a robotic device called HAL (short for Hybrid Assistive Limb but also the name of the killer AI in *2001: A Space Odyssey*). In Japan, robots from pop culture are regularly used to promote real-world robotics projects.[14]

This further highlights the fact that robots are developed in specific historical, sociocultural, and political economic circumstances. Grasping this context is critical to understanding where Japanese care robotics came from, how and why the field developed as it did, and where it is going. But this understanding requires nuance. The context in which robots are developed in Japan is clearly different from that of other countries, and robots produced there often appear to share some common design characteristics. Nevertheless, although it may be tempting to talk about a "Japanese robotics" with uniform properties based on the supposed uniqueness of a seemingly bounded and homogenous Japanese culture in opposition to an American or "Western" robot culture, or on a generation of Japanese robotics engineers all influenced in similar ways by popular cultural depictions of robots such as Astro Boy, this can also mislead. As anthropologist Kubo Akinori notes in his 2015 book *Robot Anthropology*, there is a danger of oversimplifying complex and heterogeneous practices relating to robotics when making binary comparisons between "Japan" and "Europe/America," or East versus West, selecting only elements that fit such a comparison instead of attending to the spectrum of different voices and practices both in Japan and other countries. Robotics is a highly international field that resists characterizations of clearly bounded national robot cultures, despite continuous attempts to do so; although national historical, cultural, and policy contexts are extremely important (see, e.g., Schodt 1988; Šabanović 2014; Frumer 2018), diverse global streams of influence, practice, and materials have also been, and continue to be, key to the development of robotics in Japan.

Pepper, a humanoid robot mentioned above that will be discussed in detail in chapter 6, is a pertinent example: although it was distributed by a Japanese company, it was developed by French engineers designing for the Japanese market, was manufactured by a Taiwanese company in China, and could be outfitted with apps written by programmers around the world. Paro, a seal-shaped robot that we will consider in chapter 5, is often presented as quintessentially Japanese. But it was developed by a Japanese engineer who had to travel to the Massachusetts Institute of Technology in the United States, then the center of a new approach to interactive social robotics, to build it because such social robots were not in favor among research bureaucrats in Japan at the time, and he could not secure funding to pursue the project at AIST. Images, ideas, practices, and practitioners of robotics circulate globally, and although certain types of funding and certain research agendas have predominated in Japan compared with other countries, there has also been a large degree of international interchange of ideas and technologies, which has been actively promoted by national and international research organizations. At the same time, there is significant variation in approaches in Japan itself, with some robotics engineers favoring creating humanoids that look as humanlike as possible, while others have specifically designed robots with abstracted, non-humanlike features to avoid the so-called uncanny valley effect (Mori 1970); some have developed "weak" robots that depend on humans in order to encourage a symbiotic relationship with their users (Okada 2012, 2016), while others have developed a sixty-foot-tall, twenty-five-ton Gundam robot that symbolizes superhuman strength and pop culture nostalgia.

Given their complex, open-ended nature, it is perhaps unsurprising that in approaching robots in Japan we are presented with a tangle of pop cultural representations, sociotechnical imaginaries,[15] political agendas, invented traditions, research prototypes, and everyday commodified products, often mixed together, conflated, or confused both in public and even academic discourses and representations. Media portrayals of Japanese robots often feature photogenic humanoid robots such as Honda's ASIMO, AIST's female-gendered HRP-4C, and the aesthetically lifelike android robots of Ishiguro Hiroshi. Articles on robot care in particular frequently lack context, usually focusing on short-term projects featuring a seemingly promising robot prototype rather than taking a broader view of how robots and other digital technologies are transforming or propose to transform care at the systemic level. They have often featured a photo of the iconic lifting robot Robear, which dominated Google Image search results for "care robots" in English and Japanese at time of writing, and evocatively juxtaposes industrial heft with a cute, cartoonish teddy bear face (see figure 1). In fact, none of these robots are representative of the kinds of robotic devices that are actually being implemented on a wider scale in Japan today. Robear, developed in 2015 as an experimental research project, was never used in a real care home setting, and the project has long since been retired, while its inventor has claimed that it was not a solution to the problems facing the industry in Japan and that migrant care labor was a better answer (Emont 2017). Ishiguro's android robots are expensive to produce, have been largely used for lab-based research or media appearances, and have yet to find widespread commercial application in public settings, serving rather more effectively as a techno-nationalist or techno-orientalist advert for

FIGURE 1. Robear (source: RIKEN website, https://www.riken.jp/en/news _pubs/research_news/pr/2015/20150223_2/).

Japanese ingenuity and soft power. Similarly, ASIMO was used mainly as an ambassador to show off to (often somewhat bemused) foreign leaders and international media—for example, playing soccer with an apparently underwhelmed Barack Obama during the president's state visit to Japan in 2014. Yet, its carefully choreographed routines were remotely controlled by engineers who remained out of sight, belying its image of apparent autonomy and technical ingenuity.[16] Indeed, in the field of human-robot interaction, a commonly used methodology is the Wizard of Oz technique, whereby a human operator secretly controls a robot during an experiment, while making it seem to research participants that the robot is working autonomously.[17] The metaphor of the Wizard of Oz is telling: performance, hype, and mystique underpin robotics, while actual functionality tends to trail far behind the illusion that is presented to the public.

Laboratories of the Future: Studying Places Where Reality Happens

The world of care robotics is complex and difficult to grasp from just one viewpoint—whether technological, cultural, macroeconomic, or political. A good place to start is where robots are actually developed and used day to day. This book is based on ethnographic fieldwork and interviews I conducted between 2016 and 2020 in Japan at key sites of care robot development and implementation. This included three months spent at AIST, which was in charge of the Robot Care Project mentioned above. I also spent seven months at Sakura nursing care home in Kanagawa Prefecture, southwest of Tokyo, before, during, and after the introduction of three different types of care robot for a six-week trial: Hug (a lifting robot), Paro (a communication robot shaped like a baby seal), and Pepper (a humanoid robot used for recreation sessions). These robots were chosen in conjunction with the manager of Sakura based on several criteria. The robots had to address the needs of Sakura, as defined by its manager. We wanted to look not just at one robot at a time but several used simultaneously, reflecting how the Japanese government expected them to revolutionize all aspects of institutional care work in the near future. All three were to be commercially available robots that were actively being marketed to care homes, not the one-off prototypes that were often used in heavily curated, narrowly defined, and short-term studies. All three were supported in some way by the Robot Care Project and were thus included within the government's overarching national care robot strategy.

At both research locations, I interviewed and chatted with people, including engineers, research managers, care workers, older adults, and community volunteers. I employed the anthropological methodology of "deep hanging out" (Rosaldo, cited in Clifford 1996), observing what they did every day and how they talked about it. I also draw on interviews and conversations with various other people involved in the scattered world of care robots: managers at companies producing robots and other technologies used in care work; programmers and tech support staff working in a robot atelier in Tokyo; representatives from local and national care organizations and government departments; managers at AMED; academics from a variety of disciplines researching robots; and care managers who had to decide whether to recommend robots to older adults and their families. Most of these conversations and interviews were conducted in Japanese, and extracts provided in the text as well as quotes from other Japanese-language sources are my own translations. I use pseudonyms to protect the anonymity of Sakura, its care workers and residents, as well as some robotics engineers and other interviewees. I keep, however, the real names of some researchers and managers who are public figures or speaking on the record. Chapters 4 and 6 are partly based on articles I published in the journals *Asian Anthropology* and *Critical Asian Studies*.

Using a multisited ethnographic approach can help make connections between the different scales of the macro and micro. People I spoke to during my fieldwork often talked about the *genba*, a Japanese term that literally means "places where reality happens"—the actual site or scene of an event. Genba is often used in news articles or conversations—for example, when talking about care homes where robots are being implemented. Ethnography can be understood as the study of genba—observing and paying attention to the detail of everyday lives in the specific, salient places where reality happens. Attending to everyday realities of genba can help us break through government and media narratives and national stereotypes. This is particularly important in the study of Japan, which has often been exoticized through the use of techno-orientalist stereotypes: portrayed as a nation full of futuristic technologies and either conformist, emotionless, robot-like citizens or, alternatively, fervent technophiles. It also helps us overcome the enchanting open-endedness and hype surrounding ideas and imaginaries of robots and focus on the lived realities being co-constructed by new caring assemblages involving robots, and the corporations and state actors involved in developing, promoting, and operating them. As I hope to show, the roboticization of care is not a straightforward process of placing useful devices in care homes to solve specific discrete problems and replace human care staff. Instead, it precipitates new forms and arrangements of care and care work, creating both new challenges and possibilities.

In order to grasp the context of robot care, we have to enter both the robot research institutes and nursing care homes that are involved in trying to build this emergent techno-welfare state. Decentering such research from Western cases is also significant: for example, whereas most technological development and policy discourse in Europe and the United States is focused almost exclusively on self-care and independence, the Japanese form of robot care has been predicated largely on trying to maintain current institutional care practices in high-tech form.

Although many actors are involved in the development and implementation of care robots, this book mainly focuses on two institutions and two sets of workers at opposite ends of the socioeconomic scale: AIST, with its predominantly male, highly qualified robotics engineers and programmers, and Sakura, a nursing care home with predominantly female, relatively less formally educated and credentialed care staff and volunteers. The two sides of development and usage are seldom brought together in one study, and indeed, as we will see, in Japan these two worlds rarely seem to cross paths at all. The aim of this approach is not to show direct impacts of robot design on robot use but rather to demonstrate the connections and, more often, disconnections between the two, and their significance for the future of robot care more broadly.

On their face, the two institutions seem to represent alternative Japanese futures: one forward facing, dynamic, and technologically advanced; the other aging, set in its ways and technologically conservative and backward looking. Yet despite their clear differences, both places and groups of actors shared somewhat surprising similarities. Both were highly controlled institutional environments that served as venues for research and experimentation into novel technologies and techniques of care and communication seen as vital for the future of an aging Japan. Both existed at the overlap of public and private sectors and were overseen by managers coming to terms with tightening budgetary constraints. Both, on the surface at least, seemed highly bounded and traditional Japanese institutions that had existed in a similar form for decades but were in fact permeated by global flows of information, practices, values, and staff. I came to see the care home as in some sense an extension of the research institute—an adjoining step in the chain of developing and testing robot care as well as experimenting with other communication techniques and technologies, and indeed a laboratory for ongoing scientific and philosophical research into the nature of humanity. Moreover, both the engineers working on robots at AIST and the care staff operating them at Sakura acted as intermediaries between developers (private companies) and presumed end users (older adults who required care). As will be seen, in mediating these relationships, they each constructed robot care in very different ways.

Japanese publicly funded institutional nursing homes are sites where many of the most contentious issues affecting contemporary Japan intersect, including perceived problems related to the aging population and political strategies aimed at addressing them. They are genba where shifting labor, welfare, innovation, migration, and regional development policies meet changing gender politics, care ethics, cutting-edge technological development and emergent communication techniques, as well as ideological, and practical, questions about how limited national resources for welfare should be distributed. If one way of seeing Japanese society, presented at the start of this chapter, is in statistical terms as an aging postindustrial economy, another is to see it as an assemblage of caring institutions, practices, and actors. In this sense, the care home can be viewed as a microcosm of Japanese society. Yet, this is not just a story about Japan and the positionality of robots in Japanese national identity, as other aging societies increasingly look to the country as a pioneer in care techniques, technologies, and institutional practices and arrangements. In 2019, the UK government announced an investment of £34 million ($48 million) in robots for adult social care, stating that they could "revolutionize" the care system, and highlighting Paro and Pepper as successful examples of such robots. The European Union has also invested heavily in care robots through its 2015–20 €85 million ($103 million) Robotics for Ageing Well program among others (see Lipp 2019; Wright 2021). As well as places where everyday reality happens, care homes are, like robot labs, sites of national and indeed global future making.

Care robotics is an area in which top-down approaches clash with the necessity for bottom-up implementation. Robotics research is usually expensive and tends to require sustained funding not just over years but decades in order to lead to products that can eventually be commercialized. Yet compared with the state resources thrown into development, the actual implementation and everyday use of care robots often seem an afterthought and are typically devolved to the hyperlocal, if not the individual, level. The users themselves—older adults and caregivers—similarly often seem an afterthought to developers; most care robots were never developed specifically with care work in mind—the solution came before a consideration of the problem and was adapted to fit retrospectively. This mismatch helps explain some of the difficulties faced in disseminating care robots.

The structure of this book is aimed at exploring elements of these top-down and bottom-up approaches. Chapter 1 examines the origins of Japan's current care crisis by looking at the history of care and welfare provision in the country. It considers how robots came to be seen as one of several possible solutions to the crisis by some in government, how policies were implemented to support robot care at the national and regional levels, and why it remains controversial. In chapter 2, we enter the world of AIST to address the everyday activities of robotics engineers working on the Robot Care Project in the context of ongoing institutional change. The chapter looks at how engineers approached questions of robot ethics and the role of standardization and explores how they constructed a vision of what I call algorithmic care through their development practices.

Chapter 3 switches to the genba (actual site) of care at Sakura nursing home before care robots were introduced and addresses the questions of who does care, how, and why. Chapters 4 through 6 look at robot care from the perspectives of staff using each of the three different robots in daily practices of care with nursing home residents. Each chapter considers a different robot in turn—Hug, Paro, and Pepper—and how they reconfigured three different everyday activities that were central to care and life at Sakura: lifting, communication, and recreation. Although they are very different types of robots, similarities emerged in use when viewed through the lens of existing meanings and practices of care at Sakura. I conclude in chapter 7 by bringing these threads together and thinking about what care robots mean for the future of care and welfare in Japan and beyond. As will become increasingly apparent through the course of the chapters, there are other cross-cutting structural dimensions involved that are less immediately apparent yet come to play a vital role in care robotics, drawing together the lifeworlds of engineers, researchers, care workers, and older adults living in nursing homes.

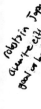

As I hope to show, the attempt to implement care robots in Japan points toward an emerging configuration of care in which human labor, far from being replaced by machines, instead becomes ever more in demand. Yet the nature of the work itself is increasingly deskilled, devalued, and alienated. Introducing commodified and interchangeable practices of robot care turns out to help render *human* care labor similarly more commodified and interchangeable, potentially extending existing forms of inequality and precarity, especially in the growing migrant care labor market. At the same time, through the process of examining stumbling attempts to roboticize care in Japan, characteristics of good care and ethical care practices are also made visible, particularly through staff resistance to new robot care devices and procedures. Such resistance is often dismissed as Luddism, or as a problem of change management or "culture," or of the tech-challenged demographics of care workers who tend to be older women. But in attending to the everyday realities of care work and the life of the care home up close, the *nagare* or "flow" comes into focus—the spatiotemporal rhythms of care profoundly linked to ethics and ideas of what constitutes good care in Japan and, perhaps, beyond. It is into the midst of this flow that robots are thrown to sink or swim.

1
CRISIS AND CARE ROBOTS

How did robots come to be seen by some government technocrats as the best solution to Japan's care crisis? At first, the answer may seem self-evident. For over three decades, Japan has been portrayed as a "robot kingdom": a kind of posthuman utopia replete with robots and other advanced technologies that are widely accepted as friends and companions due to supposedly unique characteristics of Japanese culture.[1] This is a brand of techno-orientalism that has been heavily promoted by the government and media both inside and outside the country for decades:[2] the image of a Japan where people mourn their robotic pets at Buddhist memorial services (White and Katsuno 2021), which might themselves be officiated by the humanoid robot Pepper acting as priest (Gibbs 2017), or where lonely men marry their holographic girlfriends (Jozuka 2018). But this portrayal of fervent and widespread acceptance of futuristic digital technologies fails to represent most people's everyday experience. An iconic counterexample is the ongoing widespread use of the fax machine, still an indispensable part of any Japanese office in 2020. During the early stages of the COVID-19 pandemic, the Ministry of Health, Labour, and Welfare was criticized for requiring doctors to handwrite and then fax reports of new viral infections, rather than allowing them to be sent by email.[3] Yet, 2020 was also the year in which Japan's Fugaku supercomputer was crowned the world's fastest and produced cutting-edge modeling of how the virus spread in the air. Japan has a complex relationship with technology, combining a deep conservatism with protrusions of cutting-edge innovation and communities of highly enthusiastic early adopters.

To address the question of how care robots came to be seen as the "necessary technology" that many engineers I interviewed referred to, we first need to understand how Japan's crisis of care came about, because the discourse of crisis is so central to the arguments made for robots. As we have seen, official statistics portray a future of demographic decline in Japan. New terms are invented to communicate a unique and dramatic urgency that sets the country apart: Japan is regularly described as "super-aged" or even "hyper-aged" with an "ultra-low" fertility rate. As medical anthropologist Elana Buch notes, "A[n] . . . 'apocalyptic demography' of catastrophist public and official discourses raises concern that the growing burdens of caring for an aging population are likely to devastate both families and polities" (2015, 278). Such crisis discourses have been employed in the United States and Europe to advocate for loosening immigration restrictions and to promote neoliberal-shaded agendas of "active," "positive," or "healthy" aging, which often imply that older people should be responsible for their own health and care. In relation to care technologies specifically, Louis Neven and Alexander Peine (2017) argue that a "crisis account of aging" has become a ubiquitous way of framing the relationship between aging and innovation among policymakers, scientists, and technology companies, as well as a highly effective strategy to mobilize funding for projects developing care technologies.

These types of discourse are also regularly employed in Japan. Citing attention-grabbing and anxiety-inducing demographic statistics has become a kind of ritual performance that precedes almost any written work or presentation on care robots. Japan's Cabinet Office regularly details the urgency of the situation through its *Annual White Paper on the Aging Society*, highlighting the large current and predicted future proportions of the population over sixty-five and over eighty-five years of age, the expected deficit of care workers, and the already high and growing costs of age-related welfare provision.[4] Although Japan currently has the highest percentage of people over sixty-five years of age as a proportion of its total population, other countries have an older median age (Monaco), faster aging population (South Korea), or longer life expectancy (Hong Kong), yet it is almost with pride that Japanese politicians insist that Japan has *the* oldest population in the world, an argument that rather counterintuitively feeds into the techno-orientalist trope that Japan is already living the future.

The continuous repetition of such statistics about aging helps form a simple singular public narrative that can make the crisis in *care* appear natural—the inevitable consequence of the biological process of aging rather than the result of political choices—by abstracting it from its social, political, and economic context. It can also provide a "natural" explanation, rationale, and justification for the development and implementation of technocratic solutions such as care robots, particularly in the absence of voices of many older people and their caregivers in Japanese media; we can observe similar discourses in the United States and Europe. But as we will see, there is little natural or predetermined about the crisis of care or the idea of using robots as its "solution."

Worrying about impending crises in eldercare not only has a history but indeed has been one of the main drivers for post–World War II welfare reform in Japan. Political scientist John Campbell (1992) argues that the history of Japanese care policy can be understood in terms of a succession of differing representations of aging as a social problem, although we may alternatively view it as a slow-motion unfolding of a single crisis gathering momentum over decades, driven by the ongoing neglect of reproductive labor. How exactly did Japan end up, in the 2010s, facing enormous care labor shortages and a critical mass of older people requiring care? In order to understand where the current crisis originated and where it is going, it is helpful to look at how the welfare state in Japan developed over the postwar period and particularly at the recent introduction of the national Long-Term Care Insurance system.

Eldercare and the Construction of Japan's Welfare State

Expectations about who should care have been strongly shaped by the imagined traditional form of the Japanese family, the *ie*, which was institutionalized and written into law during the Meiji period (1868–1912).[5] According to this model of the family, the eldest son would live with his parents, together with his wife and their children, in the main family house, which he alone would inherit on his father's death. He would become the primary keeper of the family altar, making daily offerings and prayers to the ancestors, and would also be legally responsible for his parents' well-being in old age. Women were expected to fulfill the role of "good wife, wise mother" (*ryōsaikenbo*), which included helping to provide or manage care for her husband's parents. Although the ie system was banned after World War II by the occupation administration, which promoted the nuclear family instead as a household unit that supposedly embodied "modern" democratic values, it continued to exert a lasting influence as an ideal type, seemingly exemplifying Confucian principles of filial piety, and serving to justify and rationalize gendered divisions of labor inside and outside the home.

As postwar family sizes gradually shrank and a growing proportion of the population came to live in increasingly closed units frequently housed in urban areas in small apartments designed specifically for one- or two-generation families, the ideal of caring for several older parents in a shared three-generation household became difficult to achieve. An insurmountable gap was developing between the social and moral ideals of familial care and the expectations and realities of modern urban living. The situation was further exacerbated by the fact that with improved nutrition and health care accompanying Japan's modernization and economic growth, life expectancies were rapidly increasing: older people were living longer and increasingly suffering from chronic diseases that required substantially greater amounts of care from family carers over longer periods. Public awareness about the challenges of caring for older relatives living with conditions like dementia grew particularly from the 1970s through portrayals in the media as well as many people's direct experience.

In line with the growing recognition of the need to provide care to an increasing number of older people and driven by Japan's booming economy, welfare services were greatly expanded from the 1970s onward. The government declared 1973 "year one of welfare" and implemented a number of new services including pension benefits and free medical care for people over seventy-five years of age. Uptake of these services rose sharply as did expenditure: local government spending on welfare for older adults grew by more than ten times between 1969 and 1974 (Campbell 1992). The year 1973 also marked the start of what sociologists describe as the second demographic transition in Japan, with the end of a total fertility rate at replacement levels. In the absence of a countervailing increase in immigration, this eventually led to annual net decline in the population, starting in 2007. The reduction in the fertility rate has been attributed to a combination of factors: the growth in women's education and labor participation, later marriage, the widespread availability of contraception, and the rising cost of childcare and inadequate provision of childcare facilities. As gender roles continued to change, a reduction in living together with older relatives tracked increasing state welfare provision (Yamato 2006).

The accelerating economic growth that paid for the expansion of the welfare state during this period was driven by export-led manufacturing that was starting widely to employ factory robots. The first industrial robot was brought to Japan by Kawasaki Aircraft (now Kawasaki Heavy Industries) in 1969, under license from US company Unimation. Whereas 1973 was "year one of welfare," 1980 was dubbed "year one of robot popularization" by the Japan Industrial Robot Association. With considerable state assistance from the Ministry of International Trade and Industry (MITI, the predecessor of the Ministry of Economy, Trade, and Industry [METI]), robots started to be implemented in large numbers, particularly in automobile and electronics factories but also in small- and medium-sized companies across manufacturing and logistics (Schodt 1988). The reprogrammability of robots enabled an unprecedented level of flexibility, meaning that a more diverse range of parts or end products could be manufactured at a lower cost by smaller companies than was previously possible. Companies were able to meet expanding domestic consumer demand, while producing variant products for different national markets, and could respond quickly to changes in customer requirements. The use of industrial robots also meant that manufacturing output could continue to grow rapidly despite severe labor shortages, without the need for importing large numbers of foreign workers as happened in many other higher income countries. The widespread introduction of industrial robots met with very little public or union resistance and few direct job losses for the same reason.[6] By the late 1980s, robots were widely used in Japanese factories, to a greater degree than in other countries; Frederik Schodt notes that in 1986, Japan had around ten times the density of industrial robots per ten thousand workers compared with the United States (1988, 15–17). Japan was also coming to be seen as a world leader in other areas of robotics research, particularly the field of humanoids, seen in the 1980s as (in Schodt's words) "highly eccentric" but since acknowledged as a crucial domain of research due to increasing recognition of the importance of embodiment in robotics and artificial intelligence, the growing attention being paid to social robots that could be used in the service industry, and the flourishing field of human-robot interaction.

After the huge increase in welfare spending in the 1970s, the Japanese government tried to rein in expenditure with cutbacks in the early 1980s. But this proved highly problematic due to the ways in which society had already changed. Peng notes,

> The number of elderly in need of care was rising as more married women went out of their home to work; the limited institutional and public care services were retrenched by welfare cutbacks; and market provisions for elder care did not exist. The care needs of the elderly and children were expanding just as the postwar welfare regime, which was structured to accommodate such needs by enabling the women to care for their family members at home, began to crumble. (2002, 419)

One result was that many older people in need of care and unable to live alone started living in hospitals despite not having a specific medical condition—a phenomenon known as social hospitalization. Whereas in 1989, just over 1 percent of over-sixty-five-year-olds were living in nursing homes, a further 3.5–4.5 percent were living in hospitals for extended periods (Campbell 1992). This was due to a combination of factors, including the shortage of long-term care facilities and stigma of familial abandonment associated with them (Bethel 1992) and the pressures that informal long-term care was placing on family members. Some measures were taken in the 1980s to try to address this issue, such as redesignating hospitals already filled with older people as "old-people hospitals" and gearing them explicitly toward long-term care. This was, however, a short-term solution; the problem called for a more fundamental reorganization of the welfare state.

A major milestone came in 1989, when a battle for tax reform culminated in Prime Minister Takeshita Noboru's fateful introduction of Japan's first consumption tax, at a seemingly modest rate of 3 percent. The imposition of the tax coincided with the peak of the twin stock market and property bubbles and triggered the bursting of Japan's bubble economy, constituting the proximal cause of the country's so-called lost decades of low economic growth in the 1990s and 2000s. Shortly afterward, the Ministry of Health and Welfare launched the Ten-Year Strategy for the Promotion of Health and Welfare Services for the Elderly (commonly known as the Gold Plan)—a major investment pledge funded by this new tax, aimed at expanding care services and significantly boosting long-term nursing care facilities. The fact that a tax to fund eldercare appeared to have dealt a knockout blow to the Japanese economy at its height associated the two in the popular imagination. The welfare of older people appeared to be inversely linked to the welfare of the nation; it began to seem that Japan may have passed its prime both economically and demographically. This perception was intensified shortly afterward by the 1990 "1.57 shock"—the news that Japan's total fertility rate had dropped to 1.57—well below the replacement level of 2.1—which was seen as an unsustainably low rate that would shrink and age the population: a further significant blow to Japan's already rattled self-image.

Some scholars have argued that the bursting of the bubble economy, in ultimately foreclosing the promise of lifetime employment as the norm (at least for men) and guarantees for the young of better lives than their parents led to a transformation in intergenerational relationships: Hashimoto Akiko and John Traphagan argue that "children's expectations regarding filial bonds shifted to more individualized, voluntary ties" (2008, 9). As anthropologist Iza Kavedžija writes, "The resentment felt by the younger generations, especially young urbanites who work hard and experience financial hardships while feeling that they are ensuring a very comfortable existence for the older generations, has led some to speak of 'intergenerational exploitation' as replacing intergenerational solidarity" (2020, 227). Economic inequality has grown during the decades of slow growth that followed 1989, with the percentage of workers in nonregular (including part-time and temporary) employment rising from 15 percent in 1982 to almost 40 percent by 2014 (Gordon 2017), a situation facilitated by a major relaxation of labor laws in the late 1990s. Yet while economic and social precarity particularly among younger people grew, older people, who form an important and reliable voting bloc for the conservative ruling Liberal Democratic Party, were about to gain access to a greatly expanded and universal eldercare system.

Long-Term Care Insurance

Even the increases in expenditure via the Gold Plan and its 1994 successor, the New Gold Plan, were not sufficient to deal with the rising demand for eldercare, and as a result, legislation to enact a more comprehensive Long-Term Care Insurance (LTCI) system was passed in 1997 and implemented in 2000. This plan, similar in structure to Japan's universal health care insurance system, introduced a new tax on people over forty years of age to help pay for universal coverage of care provision for those over sixty-five years of age and those with a disability over forty years of age. Older people in need of care services pay 10–30 percent of the cost based on a means test up to a specified ceiling, with the remainder paid by a combination of revenue from the new tax and from local and national general taxation.

LTCI represented a huge increase in the size of the Japanese welfare state. Annual expenditure on the system grew to ¥10.4 trillion ($1 trillion) by 2016 and is expected to reach ¥21 trillion ($2.1 trillion) by 2025 (Brucksch and Schultz 2018). As of 2017, 6.4 million people (about one in five of those over sixty-five years of age) had been assessed as requiring some form of long-term care or support, while the number of people living at residential eldercare facilities rose from 819,091 in 2000 to 2.13 million in 2018 (MHLW 2017a, 2019). LTCI created a new role of "care manager," a professionally qualified care worker who would carry out assessments of care need and create individualized care plans for older people. Care levels range from one (lowest) to five (highest), with two additional levels relating to long-term support rather than care. The system introduced a greater degree of choice for older people to select care services; in order to provide this choice and introduce greater competition into this new market, the government encouraged more private sector involvement in the care industry.

The imposition of a new tax marked a definitive shift toward the perception of care as a universal right for older people and away from both a reliance on family for its provision and the stigma associated with nonfamilial care. For the first time in Japan's history, the legal—and to some extent, moral—responsibility to provide care to older people in need passed from families to the state, bucking the neoliberal trend in many Western countries to try to shrink the welfare state and make individuals and families as far as possible responsible for their own care. As Yamato Reiko (2006) notes, far from being rooted in a static culture of Confucian familism, expectations of mutual reliance between older Japanese people and their children have decreased in proportion to increases in government welfare provision. The period since the introduction of LTCI has seen growing numbers of older people living alone or in residential care facilities. Although the family is still a major source of informal care, the trajectory of welfare shifted significantly toward greater state provision and treating older people not in terms of their relationship to their family, which had been the basis for most postwar welfare policy, but rather as independent individuals. The socialization of the cost of care has accompanied the growing individualization of Japanese society.

Encapsulating some of these attitudinal changes, feminist scholar Ueno Chizuko has written a series of best-selling books in Japan in which she argues passionately for the benefits of accepting and indeed embracing living alone in old age.[7] She provides a step-by-step guide for preparing to be as self-sufficient as possible in later life rather than succumbing to what she calls "the devil's whisper": the invitation to go and live with one's children, which she sees as a flawed and outdated cultural ideal that benefits the well-being of neither party (Ueno 2011). It is striking that a practice of familial care that had until fairly recently epitomized filial piety could now be regarded as positively sinful. Yet, as Joan Tronto notes in her discussion of how the neoliberal idea of freedom is imagined as a "lack of attachment," "even if we could be free from all forms of dependence, that would not be a free life, it would be a life devoid of meaning" (2013, 94). In Japan, the flipside of greater emphasis on individualism was seen in the emergence of a public discussion about *muen shakai*—literally, "a society of no ties," or a society of individuals in which traditional bonds of family and friends have broken down—spurred by a 2010 documentary by Japan's national public broadcaster NHK that highlighted the large number of older people living and dying alone, and other cases of breakdowns in social relationality.

Part of the rationale for the transformation of welfare was to free up female family members, who would otherwise be providing informal care to older relatives, to join the workforce.[8] Indeed, there was an expectation that many of these same women would be recruited into the formal care labor market. Transforming familial carers into paid care workers meant that what had previously been statistically invisible, unpaid domestic labor would start to be officially counted, thus generating "new" economic growth and contributing to gross domestic product and jobs figures.[9] The sector could be rationalized, rendering care labor a more liquid commodity: a more efficient, standardizable, commercial service transferable beyond the bounds of kinship, delocalizing care services and enabling economies of scale. But the rapid uptake of care services through LTCI created a new problem: a massive and growing labor shortage, as the reduction in reliance on familial carers has not been met by an equivalent rise in the number of paid care workers—a job characterized in Japan by low wages and social status and with a reputation for being tough both physically and emotionally. In 2020, there were over four care worker job openings per applicant, compared with one-and-a-half job openings per applicant across all professions (MHLW 2021), while the annual attrition rate in the sector is also high for Japan, at around 17 percent in 2016—itself driven by the additional stress of not having enough coworkers as well as low pay (Care Work Stabilization Center 2017, 2020). Some new entrants to the care industry have been male blue-collar workers who lost their jobs when manufacturing was outsourced to China and other countries following industry restructuring in the 1990s and 2000s. These new workers, however, have not been sufficient to fill the growing labor shortage. Although the number of formal care workers increased from 0.6 million in 2000 to 1.7 million by 2015 and is expected to reach 2.3 million by 2025 (Brucksch and Schultz 2018), this is expected to fall short of demand by 337,000 in 2025, according to an estimate by the Ministry of Health, Labour, and Welfare (2018). As a result, many care homes have entered a chronic state of critical understaffing.

In response to some of the problems caused by the fast growth in care services and soaring expense of providing them, the government continued to modify LTCI. One of the most consequential of these adjustments—and perhaps the most relevant to the introduction of care robots—was legislation that came into effect in 2015. Under this revision, older people classified at the less severe care levels one and two, who made up over half of those certified as requiring some level of care as of 2017, were no longer eligible to apply for entry into nursing care homes as they previously had been. At the same time, care homes were incentivized to accept new residents at the most severe care levels of four and five, meaning that even people at level three were less likely to be accepted into residential care facilities. The aim was to mitigate the rise in care costs due to the high price of institutionalization, but this adjustment resulted in important structural changes in the conditions of care in Japan.

On the one hand, it meant that more older people now lived at home for longer because they were not deemed dependent enough for institutional care, and many of these required more formal in-home support. This in turn has led to the development of a growing infrastructure of serviced housing, home care services, and monitoring technologies intended to enable family members to keep an eye on older relatives who live alone or who are left at home while they go to work. This is particularly the case as LTCI disincentivizes co-living with a close relative, as this can mean missing out on some financial support.

On the other hand, it meant that people entering care homes now required on average more intensive levels of care and stayed there for shorter periods, as those at higher care levels are likely to pass away much more quickly. The faster pace of resident turnover has in turn placed intensifying pressure on care homes to fill beds as quickly as possible in order to maintain their level of public funding, which is calculated per resident. Although care homes are often oversubscribed and many have long waiting lists, the process of assessing and accepting new residents can be longwinded and bureaucratic, leading to delays and empty beds.

Migrant Care Workers: METI to the Rescue?

Considering the huge rise in demand for care workers following the introduction of LTCI, two questions arise. Why have care worker wages not risen in line with demand, and why has Japan not followed the route taken by other wealthy nations particularly during the 1980s and 1990s and imported care workers from lower income countries en masse?

In response to the first question, economist Genda Yuji (2017) argues that the substantial increase in nonregular employment across Japan over the past decade has meant that labor has become extremely elastic, reducing the need for wages to keep pace with rising demand. In addition, new care and health care workers have been entering from the manufacturing sector, which has been shedding jobs as production has moved offshore to China and other low-wage countries, further offsetting the need for wage rises in care. But the tight regulation of the care sector plays the most important role: the government fixes the incomes of care homes in order to limit the overall cost of care (and of LTCI premiums), meaning that care homes operating within the LTCI framework simply cannot afford to pay care workers significantly higher wages.

The usual answer to the second question is that Japanese people and their conservative government do not want immigrants. Japanese ministers including Prime Minister Shinzō Abe had until recently publicly played down the idea of increasing levels of immigration. Japan's policies are perhaps better described as migration rather than immigration policies, since they have aimed to make the presence of most foreigners in Japan temporary, and their continued residence subject to direct control by the government. Yet, the Japanese government has long toyed with the possibility of bringing in substantially more workers, particularly from Southeast Asia, to fill gaps in the labor market and has in fact repeatedly introduced policies to try to import foreign care workers.

METI and its predecessor, MITI, have for decades been involved in trying to find economic and technological solutions to the growing demand for care, including through migration. By contrast, the Ministry of Health, Labour, and Welfare (MHLW) has broadly framed the issue as a problem that could be solved domestically through better training and incentives for Japanese care workers and nurses, arguing that the lack of caregivers was an artificially contrived labor shortage, as higher pay and improved working conditions would increase the numbers of Japanese care workers entering the profession (Świtek 2014). METI's interest in care policy can partly be explained by what Ian Holliday has described as the "productivist" nature of the welfare state in Japan, characterized by "a growth-oriented state and subordination of all aspects of state policy, including social policy, to economic/industrial objectives" (2000, 709).[10] According to this view, Japan's welfare state was developed primarily by economic rather than social policymakers as part of a broader postwar economic nationalism that continues to this day.

An early solution proposed by MITI was the 1986 Silver Columbia Plan, which suggested sending older Japanese people to Australia, Spain, and Brazil for care. The plan was quickly withdrawn following criticism that it was tantamount to abandoning the elderly abroad. Subsequently, shortly after the introduction of LTCI, METI, together with the Ministry of Foreign Affairs, took the lead in negotiating economic partnership agreements (EPAs) with the Philippines, Indonesia, and Vietnam. These agreements included provisions to allow in small numbers of qualified foreign care workers, in the teeth of opposition from MHLW and nursing organizations, particularly the Japanese Nursing Association and Japan Nursing Federation, which had for years lobbied successfully to restrict the numbers of foreign nurses and care workers entering Japan (Świtek 2016, 5–12). The result of these internal disagreements was a messy compromise. An annual quota of one thousand care workers from each bilateral partner country was introduced, but the requirement to pass a Japanese national examination at the end of the initial four-year "candidature" period meant that very few, if any, would be likely to remain in Japan—the rest would have to return to their home country. Ultimately, the policy was ambivalently promoted and implemented, proving equally unattractive to foreign care workers and Japanese care home managers, and failed to fill its modest quotas. In 2017, the number of non-Japanese people who passed the qualification exam to become a certified care worker totaled 213, compared with 65,361 Japanese in the same year (Wright 2019).

More recently, a Technical Intern Training Program (TITP), originally introduced in 1993 to bring migrant manual laborers to Japan for a maximum of three years, was extended to include care workers. This program is supposedly intended to "transfer skills, technologies, or knowledge accumulated in Japan to developing and other regions and to promote international cooperation,"[11] although it has been widely criticized for a variety of abusive and illegal labor malpractices including overworking and withholding wages from trainees and safety and labor law violations. Effective from November 2017, eldercare work was added to the scheme's roster of eligible job categories, and the length of the work visa was extended to a maximum of five years for some employers; candidates no longer required any nursing college qualification or university degree to apply. Also in 2017, rules were relaxed for international students in Japan, who could now apply for a "residential long-term care" visa and work in a care home part time during their studies. These schemes, too, had little immediate impact on the number of foreign care workers coming to Japan.

Finally, in April 2019, with mounting anxiety about labor shortages across a range of low-paid areas of employment, particularly eldercare, a major overhaul was announced: an ambitious "specified skills" work visa program was introduced with the aim of bringing 345,000 workers from China and Southeast Asia to Japan by 2025, including sixty thousand care workers (Yoshida and Murakami 2018). To put this in context, the total number of foreign workers in Japan rose from just over 0.5 million in 2009 to 1 million in 2016, and to 1.66 million by 2019; about half of this number were Chinese and Vietnamese nationals (Nippon.com 2020). The proposed strategy would continue this accelerating increase in the number of foreign workers in Japan but aim to funnel more of them into care work;[12] as with the TITP scheme, visas were limited to a five-year term. Yet over the first two years of the scheme, up to March 2021, only 1,705 people had entered Japan to become care workers using the new visa—a number that fell far short of the targets set by the government (Ito 2021). The outbreak of COVID-19 disrupted the entry of new workers in 2020, while trust in the Japanese government's treatment of foreign workers was further undermined after the country abruptly shut its borders to non-Japanese long-term residents from April 2020 for five months, trapping abroad those who had temporarily left Japan for business or holiday and separating many from their families.

These programs have so far failed to attract substantial numbers of foreign care workers. This is partly because Japan is not a particularly appealing destination for many migrant workers from Southeast Asia, many of whom experience highly stressful work environments (Asis and Carandang 2020). Trained and qualified care workers and health professionals arriving from abroad via the EPA scheme were frequently given menial manual labor tasks such as washing clothes or making beds. Anthropologist Beata Świtek (2016) has suggested that this is often intended to avoid the need for foreigners to touch Japanese bodies, but it is likely also due to a lack of understanding of or confidence in their skills and experience; foreign care workers are often seen as less professional than Japanese colleagues even when they have more formal care training, experience, and qualifications (Yamasaki 2006). The prospect of migrant care workers securing permanent residency in Japan is remote and requires overcoming daunting bureaucratic and academic hurdles: at the end of the limited term of their visa, migrant care workers have to leave Japan unless they study exceptionally hard in addition to their full-time job in order to master a relatively high level of spoken and written Japanese and pass the care qualification required to remain. Although the growing involvement of private recruitment agencies in the newly deregulated migration market suggests that the situation may begin to change, building a life in Japan remains a long-odds wager that many prospective care workers have so far been unwilling or unable to make.

The Development of Care Robots

LTCI undermined the basis of familial eldercare in Japan without mobilizing sufficient paid labor to replace it. In the yawning gap that was rapidly emerging between supply and demand, care robots appeared to have found their moment. The introduction of LTCI set the scene for national projects of care robot development by contributing not only to a national shortage of caregivers but also to a fundamental reshaping of social relations of care. Care was now increasingly seen by government at a macro level as a market commodity, composed of discrete, individualized, and diversified care services and technological goods, and as a consuming activity involving personal choice for older people. Long marginalized as consumers but now emerging as a huge and growing asset-rich market segment, they were characterized as clients or service users (*riyōsha*) rather than moral dependents. At the same time, the implementation of LTCI appeared to have established a centralized fiscal and bureaucratic framework and subsidy system that could also be used to facilitate the mass adoption of assistive devices, including care robots. LTCI, as the national marketplace into which such devices could be channeled, seemed to hold the key to the success of the national robot care strategy.

Government plans for the roboticization of care started to be developed around the same time that LTCI was being launched, via a series of research projects aimed at the development and promotion of care robots. METI has been at the forefront of working to operationalize this strategic vision, developing ideas around infrastructure, standards to support the industry, technical assistance and development funding for robotics companies, and subsidy schemes for care institutions and private individuals to lease or purchase the devices.

One of the clearest visions of a pro-robot future was presented in *Innovation 25*, a government strategy document released in 2007 during Shinzō Abe's first term as prime minister. Published first as a written document and later as a manga comic, *Innovation 25* featured a fictional account of one day in the life of the Inobe[13] family in Japan in the year 2025, in which robots were depicted performing various domestic duties that included eldercare. This was followed by the first major state-funded project for robot care, the 2009–13 ¥6 billion ($60 million) Project for the Practical Utilization of Personal Care Robots (*seikatsushien robotto jitsuyōka purojekuto*), overseen by METI via the funding agency NEDO. This project helped define the field and build some of the basic underlying conceptual and technical infrastructure that would be needed for the new generation of care robots just around the corner. One of its main outputs was the creation of an ISO standard (ISO 13482) titled "Robots and Robotic Devices—Safety Requirements for Personal Care Robots," which was implemented in 2014. Its development was led by Japanese researchers and accompanied by the construction of a large and expensive Robot Safety Testing Center designed to certify robotic care devices to this new standard.

ISO 13482 was important in solidifying ideas about care robots and will be discussed in more detail later on. Here, it is important to note that the standard centered on the idea of the "personal care robot," which it defined as a "service robot that performs actions contributing directly toward improvement in the quality of life of humans, excluding medical applications."[14] By defining care robots specifically *not* as medical devices (*iryōkiki*), this has meant that in Japan they are categorized instead as welfare equipment (*fukushiyōgu*).[15] Welfare equipment must be designated by the National Advisory Committee for Assistive Devices, associated with MHLW, in order to be eligible for rent or purchase through the LTCI system. Inclusion in this list is dependent on various strict and sometimes seemingly arbitrary physical specifications, and the price of the devices is capped, restricting the potential profit margins of any device that is added to the list (Brucksch and Schultz 2018). Few robotic care devices have been added to this list to date.

The 2009–13 project that established ISO 13482 and built the Robot Safety Testing Center was succeeded by several further projects, including a ¥5.2 billion ($52 million) 2015 Special Project to Support the Introduction of Care Robots, the ¥12.5 billion ($125 million) 2013–17 Robot Care Project, and various other smaller-scale projects. Several local prefectural and municipal projects have also supported the development and implementation of care robots as part of regional development strategies.

Contested Care

Despite these pro-robot initiatives and investments, many of my interviewees both at the National Institute of Advanced Industrial Science and Technology (AIST) and in the care industry more widely said they believed that METI and MHLW had fundamentally differing visions of the future role of robots and technology in care—and indeed, differing understandings of care itself. Whereas

MHLW and nursing organizations have seen the challenge of providing care in terms of the need to train more Japanese caregivers and maintain a high quality of care, defined in part as care by and for Japanese people, METI has approached the issue as an economic and technological problem. This tension was described by a senior manager from a large Japanese care home company that was collaborating with several technology firms to develop and implement a range of robotic assistive devices. The company had worked with AIST, METI, and MHLW on the Robot Care Project and other related care technology projects and seemed a model partner for the government's schemes. Yet, the manager told me,

> METI doesn't really know about care. They just help with research and so on, but the ones who actually deal with practical implementation at the actual site [genba] are MHLW. MHLW are not very positive about robots—that's how it is in Japan. Why? Because no matter how hard this side tries, they don't really understand the other side [he gestures with his hands to indicate the two ministries].

An interviewee who worked in a section of MHLW involved with long-term care told me that although the ministry was not exactly opposed to care robots, it maintained a highly cautious attitude due to the cost they potentially entailed and was concerned that widespread adoption could push up LTCI premiums (the tax for those over forty years of age) to unacceptably high levels. Another interviewee directly involved in the Robot Care Project described a feeling of somebody putting their foot down on the accelerator while someone else was applying the brakes.

The chain of government-backed care robot projects seemed to have drawn a smooth straight line from strategic vision to development to promotion to implementation. The centralization of care brought about by the LTCI system and the care shortage it contributed to seemed to offer a unique opportunity to disseminate care robots nationwide. But despite domestic and international media narratives about the necessity of robots as the solution to Japan's care crisis, and about their presumed naturalness and "acceptance" in Japan, and despite MHLW having gone through the motions of supporting these projects, the senior manager quoted above added, "The actual circumstances in Japan are that it's not going very well for robot care." In order to understand METI and AIST's vision of the future of care and what made it seemingly so incompatible with the understanding of those at the genba (actual site) of care, we now turn to the team of engineers responsible for the administration of the Robot Care Project.

2
DEVELOPING ROBOTS AND DESIGNING ALGORITHMIC CARE

I arrived at Japan's National Institute for Advanced Industrial Science and Technology (AIST) in early 2016 for a three-month fieldwork visit. AIST was the main research institute leading the government's flagship national Robot Care Project. The bucolic campus in Tsukuba, a small city in Ibaraki Prefecture, was a world away from the hustle and bustle of Tokyo, just forty-five minutes south by train. Before the start of my fieldwork, I had expected to find a team of engineers, spanners and screwdrivers at the ready, collaboratively tinkering with machines in something like an auto repair shop—a warehouse-like space of oil slicks and spare parts similar to another robot lab I had seen at the University of Hong Kong. What I observed at AIST, however, was quite different. Inside its spacious, leafy compound, the buildings were dark and silent, a maze of empty corridors and almost-windowless rooms in which predominantly male engineers and programmers sat at office cubicles, quietly tapping away on computer keyboards.

Despite the apparent isolation of the team at AIST, this research center constituted the central node in a sprawling national assemblage of actors involved in the Robot Care Project. Having looked at robot care strategy in Japan at the policy level and its relationship to the historical development and current context of eldercare, we now turn to how this strategy was implemented on the ground through the daily lives and work of engineers. This chapter considers the genba (actual site) of the research institute involved in administering and supporting the Robot Care Project, and asks, what did engineers involved in the project actually do on a day-to-day basis, and how did they understand their work? What were the processes through which they attempted to roboticize eldercare?

Tsukuba, AIST, and the Robot Care Project

AIST is one of Japan's largest and best-known public research institutes, with a global reputation for innovation. It occupies an intermediary position between the government's national science and technology strategy, academic research, and high-tech private industry. Its former president served as the head of Shinzō Abe's Robot Revolution Realization Council, and its president during my fieldwork was a former head of Sony. AIST also operates as a gateway to the rest of the world in terms of research collaborations, international technology transfer, and furthering the commercial and export interests of Japanese companies across a range of high-tech industries. The organization employs approximately two thousand researchers across ten research locations around Japan, although most are stationed at its headquarters in Tsukuba.

Tsukuba in its current form was constructed in the 1960s and 1970s, during Japan's period of rapid economic growth, as a "science city," and AIST, which had been freshly restructured and rebranded, moved there in 1980. The city came to international prominence with the 1985 Tsukuba Expo and is now, according to its website, "the largest science technology accumulation site [in] the country," hosting over three hundred public and private institutions and enterprises, as well as the science, technology, engineering, and mathematics–focused Tsukuba University, which was relocated from Tokyo in 1973 and supplies graduates to many of these organizations. The city, as well as the wider prefecture, has a higher percentage of foreign residents than most other parts of Japan—many of them researchers visiting or employed by its various research institutions. The completion of a fast-speed rail line to Tokyo's tech center Akihabara in 2005 significantly reduced the travel time between Tsukuba and Tokyo, which is now a shorter journey than to the prefectural capital, Mito, further signaling its identity as more of an international research-oriented suburb of Tokyo than a city of Ibaraki Prefecture. Aside from a nearby mountain (Mount Tsukuba), there are very few sites that establish a local identity outside of the numerous science and technology institutes and facilities; Tsukuba has the oddly decontextualized feeling of a modernist planned city set down in the countryside. The city itself is an integral part of Japan's strategy for achieving economic growth and post-bubble revitalization through techno-scientific innovation.

Tsukuba is promoted as a robot-friendly city: from the naming of the local basketball team (the Tsukuba Robots) to its official designation as a special robot industry zone. This designation permitted the use of robotic vehicles at various public locations around town, and there were several robot battery-charging stations dotted around Tsukuba's train station. The city also hosts regular public events related to robotics. Although during my three months living there, I never saw a robotic device being used or charged in one of these testing areas, researchers at AIST told me there was a "robot season" every summer, when robots could be seen regularly on the street being tested in the "real world." The timing of this season was the result of funding cycles: government funds were disbursed on a particular date each year, so prototypes tended to appear at around the same time and were tested outdoors when the weather improved after the rainy season, at the end of July. The robot calendar continued in fall with the participation of companies, government agencies, research institutes, and universities in a series of robot-related trade exhibitions and promotional events in Tokyo's Odaiba district and elsewhere.

The Robot Innovation Research Center (RIRC), where I conducted my fieldwork, was made up of fifteen permanent full-time employees (robotics engineers and researchers) on loan from the larger Intelligent Systems research institute within AIST, as well as several programmers and administrative contract staff.[1] Dr. Hirukawa Hirohisa, head of Intelligent Systems, was in overall charge of the team; his deputy was Dr. Ōba. Direct interaction between the managers and team members was rare, and their offices were located in another part of the building, with operational day-to-day management of the team overseen by the team leader, Professor Matsumoto. At the time of my fieldwork, RIRC operated from an anonymous-looking, sprawling, and rather gloomy building on the AIST Tsukuba campus, and the team occupied several open-office spaces separated into cubicles. Occasional experiments were carried out in a suite of four large adjoining rooms that contained several different models of robotic devices, including a Pepper humanoid robot, two lifelike Actroid F androids, a doll-like robot, and a couple of electronically adjustable hospital beds and wheelchairs.

AIST is a focal point of international flows of information, ideas, and personnel, hosting many foreign researchers on a temporary and sometimes permanent basis, often as part of international collaborations with foreign research institutions or companies. Several Japanese robotics engineers at RIRC had spent time at research institutes abroad, particularly in the United States, and most of the engineers participated in international conferences and wrote papers in English for international journals. A number of researchers from the United States, Europe, and Australia were working there during my visit, and the team was also engaged with international standards and robotics organizations. The ideas and practices underpinning the research undertaken at RIRC did not straightforwardly constitute a spatially or temporally bounded "Japanese" robotics but rather a robotics that was, though emergent within Japanese sociocultural contexts, at the same time porous and internationally engaged.

The pervasive silence of the lab, with programmers and engineers working intently at their computers for much of the day, inhibited casual conversation and created methodological challenges to carrying out ethnographic fieldwork, limiting much discussion to the more formal setting of semistructured interviews.[2] This was exacerbated by a further emphasis on secrecy surrounding ongoing research—a fairly typical concern across the wider robotics industry. Team members were instructed to limit their conversations with me to details of the Robot Care Project which could be safely shared and to avoid discussing new projects that were planned or details of ongoing funding applications and research collaborations.

This working culture of silence seemed to have the effect of stifling debate and foreclosing resistance or dissent relating to the direction of research. This extended beyond the researchers themselves to the wider project: one team member told me that different care home associations sometimes disagreed with each other, so *all* care home associations were simply excluded from the consortium involved in the Robot Care Project. Instead, individual care homes were approached for cooperation with the project on a case-by-case basis. The appearance of consensus was important to the project's political success, but achieving it meant shutting out potentially discordant or critical voices—not least those of the very people and organizations who would be expected to use the products under development. Perhaps as a result of this workplace culture, some researchers seemed to treat my interviews as an outlet to unload some of their anxieties about the project and were surprisingly frank and critical of the management, development practices, and overall aims of the project, indicating that silence did not necessarily equate to harmonious consensus.

The objectives of the Robot Care Project that RIRC was tasked to implement were ambitious. They included supporting the development of Japan's care robotics industry by providing various types of research and testing support and bolstering the care industry by overseeing the creation of a wide range of safe and effective robotic devices to assist older people and their caregivers, ultimately reducing the public burden of care costs. These overarching objectives would be operationalized by supporting the development of eight different categories of robotic care devices. These included wearable transfer aids, which can be worn by a caregiver to help them transfer a care recipient into and out of their bed or wheelchair (e.g., Cyberdyne's HAL power suit); nonwearable transfer aids (e.g., Fuji Machine Manufacturing's Hug, discussed in chapter 4, or Panasonic's Resyone, a bed that folds electronically into a wheelchair); indoor and outdoor mobility aids (e.g., RT Works' RT2 walker, which can automatically brake when pushed down a slope and drive itself when going uphill); toilet and bathing aids (e.g., high-tech mobile toilets and electrically operated devices to transfer people safely into a bath); and monitoring (*mimamori*) systems designed for care homes or private residences. These monitoring systems are commonly described in the United States and Europe as telecare devices and use cameras or sensors to automatically detect when someone gets out of bed at night or falls over when they are alone and alert a caregiver. In 2016, a ninth category of "communication robots" was added, including already-developed devices such as Intelligent System's Paro and SoftBank Robotics' Pepper (discussed in chapters 5 and 6, respectively), PIP&WiZ's Kabochan, Sony's AIBO, and others.

Reflecting this sprawling and multifaceted scope, the Robot Care Project involved a complex, loose network of many different actors, including the RIRC team that ran the project; the Ministry of Economy, Trade and Industry (METI, the ministry that controls AIST); the Japan Agency for Medical Research and Development (AMED, the funding agency that took over supervision of the project in 2015); the Robot Safety Testing Center; an alphabet soup of industry bodies and standards agencies including JARA, JASPA, JQA, JSA, JASPEC, JNIOSH, ATA,[3] and the International Organization for Standardization (ISO); forty-eight companies involved in developing the robotic devices; and around ten local care companies and institutions helping to trial the devices. Overall, Matsumoto told me that more than twenty permanent staff and ten temporary technical staff from RIRC were involved in the project. Including all other actors from public institutions and agencies (but excluding those from private companies), around a hundred people were working on the project in some capacity, with many of these representing their own sets of stakeholders. The connections between this bewildering array of institutions and actors were managed by the RIRC team through regular face-to-face meetings and contact via email.

The Changing Face of Public Care Robotics

The members of the research center were highly qualified and experienced mechanical and electrical engineers, so it was striking that the vast majority of their research activities did not directly involve physical engagement with robots. Typical day-to-day tasks included quantitative analysis of the activities of daily life (ADLs) of older people; cost-benefit analysis to prove the business case for robotic devices to care companies and institutions; the creation of software tools to help robot companies in their development processes; devising criteria for evaluating the effectiveness of robotic devices; giving media interviews; and various other types of market analysis and research support that were provided to care robot companies participating in the project. With their work encompassing market research, academic research, business pitch, audit, and media-savvy branding exercise, team members at RIRC often seemed as much entrepreneurs as engineers.

Indeed, research entrepreneurship was the desired outcome of larger changes afoot at the institute. AIST was in the midst of a transition toward greater reliance on self-generated income through commercial activity. Matsumoto told me that the amount of public funding provided to AIST by the government had been reducing in recent years by around 1–2 percent per year so that by 2016, around half of its income came from the private sector. Although nominally a national institute, it is technically an "independent administrative agency," having been ostensibly separated from METI in 2001, partly to give the appearance of reducing the size of Japan's government and the number of government employees, although effectively it remained under METI control.

Earlier state-funded work in public research institutes and universities in Japan had largely failed to produce widely commercialized robotic products. By the early 2000s, robot research was coming to be seen even by some of those working in the field as a disconnected set of self-indulgent pet projects that wasted taxpayers' money and never seemed to lead anywhere useful (Wagner 2013, 277–83). AIST itself had been involved in a number of high-profile research debacles in the 1980s and 1990s, including an eye-wateringly expensive decade-long Fifth-Generation Computer Systems project that came to be widely viewed as a failure that contributed to an "AI winter" in Japan.[4] Partly as a result of these problems, as well as in response to a growing American focus on commercialized technoscience, new ways of industrializing AIST's research results were explored to figure out more efficient routes to commercially successful product development.

From the early 2000s, the Japanese government began to refocus public research institutes toward supporting industry and helping develop a regulatory environment conducive to private companies rather than developing robotic products themselves. As with the introduction of the Long-Term Care Insurance system in the eldercare industry, this reflected a broader ideological approach aimed at combining public and private through the creation of a marketplace closely regulated, supported, and partly funded by the state but populated by the private sector. This involved an emphasis on paid collaborations with private industry to help cover the growing shortfall in public funding—what one RIRC researcher described to me as "research for hire" on the model of Germany's Fraunhofer Institute, another of the world's top technology research institutes. The approach was particularly championed by Chūbachi Ryōji, a former president of Sony who was appointed head of AIST in 2013.

With this new strategic direction for AIST, the leadership saw the institute serving as a model and catalyst for greater entrepreneurship not just among research institutes in Japan but also across the robotics industry more widely. Corporate risk aversion had been acknowledged as a problem in the domestic robotics sector for over a decade: the 2001 JARA/JMF industry report mentioned earlier identified a "lack of entrepreneurial spirit" as the main obstacle to achieving their goal of a "robot society" (the desire of Japanese citizens for a robot society was taken for granted). Awakening this entrepreneurial spirit, however, was proving a considerable challenge, particularly since AIST could not exert any direct control over companies—and indeed, it seemed that although larger companies were more able to afford such collaborations, they were also less likely to take risks on developing new products or sharing information with outside organizations. An AIST engineer who worked with various Japanese technology companies told me,

> Whereas in the United States, tech companies throw everything at the wall and see what sticks, Japanese ones will develop everything to throw at the wall and then not actually throw it! They just assume it's not going to stick.... They just never actually try to bring them to market—either [through] fear of producing something that may cause problems or because they don't feel that the market wants it—the marketing people say no even when everyone else says yes.

One of the ways AIST was to become a model of entrepreneurship was through the creation of spin-off companies, based on the model of U.S. institutions such as the Massachusetts Institute of Technology (MIT). Matsumoto suggested, however, that these attempts had largely failed because "researchers usually are not good at making money.... We usually don't have that kind of talent." Not having "that kind of talent," which he identified as entrepreneurialism, was the basic problem identified by many at the institute. A few senior managers and researchers at RIRC did seem to personify the so-called triple helix of industry, academia, and government. For example, the head of Intelligent Systems at AIST, Hirukawa, was one of the most senior figures in the robotics world in Japan. He had taught at the University of Tsukuba, served on various governmental funding bodies and committees on robotics, and was later put in charge of the robotics showcase at the 2020 Tokyo Olympics. He was also chief technology officer (CTO) of the aRbot Corporation, a "regulatory robotics company." The chief executive officer of this company was Professor Honda Yukio, himself former head of Panasonic's robot development center, professor of engineering at Osaka Institute of Technology, and also a program director at AMED, the new agency overseeing the Robot Care Project. Dr. Shibata Takanori worked at the Intelligent Systems research institute while also serving as CTO of Intelligent System Company, which sold his invention Paro, and as lecturer at the Tokyo Institute of Technology. Yet, the success of these endeavors did not come close to that of spin-off companies in the United States, such as iRobot, Boston Dynamics, and Rethink Robotics (which all originated from MIT), and they were in any case the exception rather than the norm.

Although the squeeze on public funding for AIST may not have stimulated an explosion of entrepreneurship, it did have the effect of increasing the need for engineers to write funding applications and progress reports—a typical phenomenon associated with international neoliberal audit cultures involving the performance of accountability in publicly funded institutions. Matsumoto told me that researchers increasingly had to put together big project bids, often framed as a large-scale collaboration between the government and private sector, to act as a kind of financial umbrella under which their individual research interests and side projects could also be funded and sustained. During my fieldwork, this large amalgamating umbrella project was the Robot Care Project. Rather than representing a national project of consensus, with researchers working together on a single unified vision of the future of Japanese care, in reality the Robot Care Project involved researchers working as individuals on discrete, often unconnected, aspects of the project while shoehorning in their own, often tenuously related, personal research interests. For example, one researcher was developing a third generation of visual markers (after bar codes and QR codes) that could be read by robotic devices and was being trialed on the International Space Station via a collaboration with Japan's space agency, JAXA, which had a research campus close to AIST's in Tsukuba. The researcher was looking at how to apply this to the Robot Care Project, but as was the case with many personal research projects, there were no concrete plans yet for its terrestrial application.

Communication at RIRC

The complexity and difficulties of communication and social interactions I observed at RIRC contributed to an atmosphere of isolation in the institute and served as a sharp contrast with the designed simplicity, contextlessness, and naivete of the communication robots that came to be included in the Robot Care Project. "Social" robots like Paro or AIBO react to physical stimuli from the user simply and fairly predictably with gentle physical movements and nonverbal noises, and even those that talk, like Pepper or Kabochan, tend to do so using simple phrases.

Most members of the team worked independently and in silence on isolated activities and rarely seemed to communicate or collaborate on their research despite the shared umbrella of the Robot Care Project, creating a sense of alienation. Several people complained about communication in the team: "It's a problem. Currently I've worked on this project alone"; "We don't really talk"; "[We] don't really work as a team." Perhaps the most strident criticism came from a non-Japanese team member who had previous experience of working for a Japanese company and at other international research institutes. He argued that although the quality of research at AIST was of a similar level, the lack of communication and effective knowledge management made it far less efficient and slowed down development:

> Here, it's... you are as isolated as possible! [There is] no interaction! [laughs] The problem is that there is no honest communication. So for me, it's obvious. I have no resources. I don't know how to do this. I need training. I need contact with an expert. They won't tell you this. They will contact you the day before the deadline, when there is no option, and just [say], "Help me." So this is a problem.... I have a review each week—I call it iteration review—a review of the last week [sighs]. But this is strange—it's all in Japanese, and I don't understand even a single word!

Almost all the people I interviewed at RIRC complained of a lack of communication from management—including the managers themselves. Senior managers were viewed as aloof and were spatially remote and rarely seen, although they explained that this arrangement was necessary because of the growing number of bureaucratic duties they were now required to fulfill. One engineer, Nakamura, described the need for "self-management"—not only in the sense of researchers managing their own careers but also in day-to-day tasks. At the same time, access to "customers"—companies collaborating with RIRC on the Robot Care Project—was heavily mediated. As one interviewee said, "You go to your boss to call the meeting. But before that, you have to agree the agenda for that meeting because everything is confidential, so you have barriers and obstacles everywhere."

One particularly egregious example of the difficulties of communication at RIRC stood out. Dr. Eguchi, a medical doctor and senior researcher, was the sole person in charge of the communication robots' element of the overall project. Until nearly the end of my three months at AIST, I was unaware of her existence because nobody talked about her, she did not appear to be mentioned on the team website, and I never saw her interacting with any project members or coming to meetings, lunches, or the team's occasional social gatherings. She was situated in her own office separate from the rest of the team, and even the overall head of the project, Hirukawa, told me that he did not know what her plan was for that part of the project. When I asked another researcher about Eguchi's work, I was told, "Hmm, actually we don't know; even *we* don't know [laughs]. Isn't that strange? We should know that!" My inability to communicate with this researcher and arrange an interview with her was somewhat frustrating. When I sent Eguchi an email to request an interview, I received a one-line rejection and no further replies to follow-up emails. Several researchers on the project told me that she was isolated in her own room because she did not get on well with colleagues, and nobody I spoke to could help me to arrange an interview with her. Yamaguchi, a RIRC team member, told me that she was very "strict" with those working under her and that it was ironic that she specialized in communication and therapeutic "healing" (*iyashi*) robots like Paro. Yamaguchi joked, "We should ask [Paro], 'Please heal Dr. Eguchi!'" An email on this subject from a research manager was typical: "I think it is difficult to interview Dr. Eguchi. Recently, Dr. Eguchi is keeping everything secret, and nobody even in AIST can interview her about the communication robot part of the project."

Although what in the United States or Europe might be characterized as a minimal communication environment is not uncommon in Japanese workplaces, the complex and sometimes strained communication between colleagues at RIRC contrasted strikingly with the goal of creating robots intended to facilitate communication among their intended users, from whom the engineers seemed almost entirely disconnected. I experienced this disconnection firsthand during a visit with members of the team to a public nursing home for older adults situated near Tsukuba that was trialing devices developed through the Robot Care Project. I was told that RIRC made a visit to this home once or twice a year. The home had a female manager, but when we arrived, we first conducted a meeting with five male staff members; on the RIRC side, there were also five male researchers. After a discussion of the robot trial, the care staff took us on a tour of the home. We were taken to see each device in situ, including electronic walkers, monitoring systems, rehabilitation equipment, and a small communication robot called Palro. During the entire visit, I did not observe anyone from the RIRC group talk to any of the residents—the visit focused purely on the machines. We were led from device to device, with the researchers photographing each one and asking care staff questions about them.

When I later asked Nakamura, the researcher who organized the visit and acted as the liaison with local care homes involved in the project, about the lack of direct contact with any of the residents at the home during the visit, he answered,

> I haven't made direct communication with such elderly people probably. When I visit facilities, I sometimes talk with the elderly people, but I haven't had the chance to talk with the person who uses these technologies. Because I think just talking about such technologies, they tend to say [either] too good or too bad things.

I interpreted this to mean that he thought that older adult users would either praise or dismiss the new technologies outright, with little nuance in between: he went on to explain that members of the team felt they would not receive useful answers from the residents, many of whom suffered from some form of dementia. This statement was echoed in remarks by other colleagues. When I asked one senior engineer, Imai, "What kind of connections can social robots make with older people?" he paused for a long time before replying, "Honestly, in relation to this, I have no idea. I'm not of the age to consider myself elderly yet, so I don't know to what extent old people actually communicate in a satisfying way with robots."

Sociologist Raelene Wilding (2018) has argued that developers often inscribe into eldercare technologies an idea of individualism that describes themselves rather than their intended users. She argues that the societal ideal of individualism—characterized by autonomy and a lack of reliance on others—is often easier for the developers of such devices to attain because they (like the engineers at AIST) tend to be male, able-bodied, with higher salaries, and more globally mobile, career-driven, and highly credentialed. At the same time, it is often more difficult for women, older people, the less able, those with less economic or social capital, fewer credentials, and a lower-status job to attain this same ideal. The way team members at AIST communicated and worked as individuals certainly seemed an expression of this kind of autonomy. It also reflected the kind of design paternalism, often based on highly negative views of old age and of the capabilities of older people, that Louis Neven and Alexander Peine (2017) have argued is common in the development of care technologies, particularly among less diverse teams that have little substantial interaction with end users.

My aim here is not to suggest any kind of individual moral failing among roboticists at AIST. All of the engineers I interviewed told me they wanted to improve the lives of older people and caregivers through the application of technology. Indeed, many of them had devoted much of their working lives to this goal in one form or another, and there was genuine excitement about the new capabilities that robotics and other digital technologies could offer particularly to people living with severe disabilities. It is, however, important to consider how the institutional research environment and culture at AIST—with its top-down planning and bureaucracy, lack of interaction with prospective end users, and few opportunities to communicate with colleagues or critically discuss the broader project—shaped and constrained the work individual team members did as well as the manner in which they did it, and how this in turn shaped the outcomes of the Robot Care Project.[5]

Robot Ethics at RIRC

During my fieldwork at RIRC, two foreign robotics engineers—one from Southeast Europe, the other from the United States—were also visiting. Our frequent chats would almost inevitably turn to the ethics or perceived ethical implications of robots. After Boston Dynamics released a promotional video of a man harassing its humanoid Atlas robot with a hockey stick to demonstrate its ability to maintain balance, they talked about the negative public relations repercussions against the robotics industry for seeming to treat robots unethically.[6] We discussed the ethics of conducting experiments to test gaze tracking technology on monkeys. One of the researchers conducted an experiment involving a game of "blink killer" with the two Actroid F androids in which I participated and in which one of the androids would "kill" the human participants by blinking at them while making eye contact. This led to extensive debate about how this experiment would be perceived and, the engineer argued, hyperbolized by news media as further evidence of the development of "killer robots." Eventually, the two researchers stopped talking to each other after a heated argument about the ethics of military robots—an argument exacerbated by the facts that the U.S. researcher was visiting from an institute partly funded by the military (a common source of financing for robotics research in the United States), and that the European researcher had lived as a child in a city that had come under bombardment by American armed forces. The debate provided a particularly visceral example of the gulf between discussing ethics from a privileged and distanced abstract perspective and from the perspective of lived experience at the receiving end of innovative high-tech products of research institutions.[7]

Robot ethics was never a distant subject for these researchers. In fact, it was mentioned almost compulsively, indicating a growing importance to Euro-American robotics engineers also reflected in a recent proliferation of academic literature on the subject. Yet over the three months of my fieldwork and during subsequent visits, I never heard the topic of ethics raised explicitly by Japanese researchers without prompting. The low-communication environment at RIRC seemed to reduce opportunities to discuss broader topics such as questions of ethics; everyone appeared to keep their views to themselves. One of the managers of the research center told me that ethics was rarely discussed seriously among robotics engineers in Japan: "They have this kind of ethical debate abroad, but if you have this at a Japanese robotics society, probably everyone would say, 'Are you feeling OK?' or 'Are you stupid?' [laughs] So you can't really do that in Japan." In terms of the ethics of communication robots, he told me, "There's no ethical research [in Japan]. . . . The only way the communication robots are being evaluated is just on effect. Ethics is completely left to the local level—for example, if it's in a hospital, it's left to the hospital to do an ethical review. There's no argument [here] about ethics."

In an interview with the *Wall Street Journal*, Patrick Lin, one of the editors of a recent major book on the subject, *Robot Ethics*, commented that despite wanting to include a paper from a Japanese scholar or industrial expert, he had been unable to find one (Kageki 2012). Sociologist Nakada Makoto argues that there is

> limited interest in the "abstract" discussions as well as in straightforward emotional expressions with regard to robots and ICTs [information and communication technologies].... Japanese people including myself have difficulty in understanding why some of the main topics, i.e. "autonomy," "responsibility (of robot, or of artificial agent)" (and the topic "robot and ethics" itself) in robotics and roboethics in "Western" culture(s) are so eagerly discussed by scholars and authors in "Western" culture(s). (2010, 300–3)

Similarly, Kitano Naho states, "In Japan, the direction of such discussions is more practical than theoretical/philosophical" (2006, 79). Others disagree. Okamoto Shinpei (2013) argues that despite the misconception that Japanese researchers "maintain their silence on issues of robot ethics," there has actually been a long history of Japanese philosophical engagement in the subject since the 1970s, albeit centered on different concerns from Euro-American counterparts, with a greater focus on questions such as whether a robot can be designed with *kokoro* (a concept encompassing elements of "mind," "heart," and "spirit").[8] Ishihara Kohji (2014) has also written about the development of a "synthetic" approach to robotics among several leading Japanese engineers and how this has shaped ethical perspectives on their work.

When I raised the topic of ethics with Japanese researchers at RIRC, they tended to talk around it in one of two indirect ways. Despite Nakada's claim that Japanese people are uninterested in ethical abstractions, engineers talked about their obligation to help solve the problems of the aging population—a generalized abstraction of age and older people—and safeguard the future of Japan. Whether they agreed with the project's research approach or not, interviewees adopted the language of necessity, telling me in serious, emphatic tones that the Robot Care Project was "very important" and "vital" to Japan and its future. An often-repeated word was *yarushikanai*—"[we] just have to do it."

The second way in which the topic was discussed was through discourses of product safety and risk, often described in negative terms. This point was made to me most clearly by Yamamoto, a male engineer who had worked at AIST for around seven years:

Y: My impression of this [Robot Care] Project? . . . This project is very important for Japanese society and for the robotics industry or robotic society. So we *must* make a success of this project. But . . . my impression is that it's very difficult in terms of the technological viewpoint. The main reason of the difficulty is safety and cost. These two factors are very important and severe for us. My standpoint is, robots can do nothing [laughs]. Many people think robots can do everything, but many robotic researchers think that the robot can do nothing, especially in tasks involving physical contact with a human or object. Tasks involving physical contact are very difficult in terms of safety.

JW: What's the main problem regarding safety?

Y: I think robots need more experience in the real world. Such experience improves the safety and reliability of robots. But robots cannot do such an experiment [to gain] experience because of safety problems. It's a negative spiral. Mechanical and sensing, and recognition of the situation, and prediction of human emotion. Every item is insufficient by current technology.

JW: Why do you think the project is very important in Japan specifically?

Y: The problem of the low birth rate and aging society—it's a very big problem in Japan. Across the world, the low birth rate and aging is continuing—Japan is the top country for that. One solution is robotic care.

JW: Do you think it's the best solution?

Y: [Laughs slightly nervously, pauses, then continues in a quiet voice] No, no, no, no, no . . .

JW: What's a better solution?

Y: The solution is technology helping, supporting a human caregiver. It's the best solution. So the care receiver with care robot is not good. Care receiver, caregiver, and *supporting* robot—this is the best solution for robot usage. But the ultimate, ultimate target [of the Robot Care Project] is a care receiver with a complete nursing care robot—directly [laughs]. The ultimate target. It's a science fiction world. But the ultimate target is that.

Similarly, when asked about the ethics of the Robot Care Project, another engineer replied, "That's a very difficult research question! This is a really important project because we're facing a really aging society. We need something for supporting our future, so I understand that's a really important project. But I don't know whether robots are a practical solution."

For the Japanese engineers, then, beyond a general sense of helping the nation, ethical questions about care robots were reduced primarily to issues of physical risk, while mentions of social, psychological, or political dimensions of robot use were absent. Academic work on the subject, whether in Japan or elsewhere, seemed to have had little impact. As physical safety has become one of the key public discourses about robot ethics in Japan, it has simplified complex ethical issues, disregarded the politics inherently embodied in technology, and claimed an ethical position while bypassing debates around broader responsibilities associated with the research, development, and implementation of care robots.

Standardizing Robots, Standardizing Ethics

As concerns about robot ethics were understood at AIST primarily as matters of engineerable safety, so too were they translated into questions of standardization. Standardization in turn is closely entangled with industrial strategy and investment as well as international engineering diplomacy. Indeed, the government's overarching Fifth Science and Technology Basic Plan for 2016–21 explicitly mentioned the "vital" importance of "strategic use of intellectual property and international standardization in order to rapidly convert the fruits of research into commercialization and international deployment, and to improve competitiveness" (Government of Japan 2016, 19, 52–53). With Japan's robotics industry seemingly preoccupied with safety, risk reduction, and its domestic and international reputation for these qualities, one of the key goals of METI was to create new ISO standards. ISO 13482 for personal care robots had been developed by Japanese engineers during a previous 2009–13 AIST care robot project and implemented in 2014. A further standard was also under development during the period of my fieldwork, and in 2019, AIST announced the release of new Japanese (JIS) standard Y1001 relating to the operation of robot services, which they proposed to be turned into an international standard in a submission to ISO in 2020. Managers at RIRC saw these standards as a vital part of Japan's strategy for developing a global market for robot care because they believed they would facilitate the export of robotic care devices by removing legal barriers to their importation and purchase by the national health services and social insurance schemes that acted as gatekeepers to the lucrative aging markets, particularly Nordic welfare states, targeted by many Japanese care technology companies. Standards would enable both the ethical certification and global scalability of robot care.

A new ISO standard is put together by one or more technical committees (TCs) in a process that typically takes at least five years. The relevant TC for personal care robots (TC299) oversees several smaller working groups made up of expert representatives nominated by ISO member states with an interest in the standards that they will oversee; in the case of TC299 this included China, South Korea, Japan, and Germany—all countries with large domestic robotics industries and aspirations of global leadership in this emerging field. Taylor, an engineer based at RIRC, was the Japanese representative on a TC and several working groups relating to robots. He explained that in the case of Japan, the government had a standardization office through which all ISO efforts were coordinated but that robotics standardization strategy was delegated to JARA (made up of representatives from the robotics industry), ensuring strong private sector influence.[9]

Standards are not just aimed at ensuring safety and quality but equally serve as tools of national industrial reputation and branding and of international trade. As a result, the creation of standards can be intensely political. Taylor, the Japanese representative mentioned above, told me, "When I was invited to join, I thought, 'Well, this is great, I can contribute to something people are really going to use and rely on,' and then I found out just how political it is and it got . . . really depressing." Two international blocs dominated the TCs related to robotics: one led by Japan and Germany, the other by China and South Korea. Japan and Germany, as nations with large and established robot manufacturers and markets, were seeking to maintain the competitive advantage of their robotics industries by resisting efforts by China and South Korea, who wanted to break into the market, to introduce new ISO decisions on issues such as modularity. Modularity refers to the ability to standardize individual parts of robotic devices, enabling manufacturers in places like China to produce standard interoperable parts of a robot even though they may lack the ability to manufacture the whole robot. Many on the Japanese and German side, on the other hand, wanted to maintain the current proprietary regime whereby individual system components varied according to manufacturer, thereby hard-coding an existing competitive advantage into their devices.

The risk, however, of the latter approach, which an engineer at AIST characterized as "basically protectionism," is a "Galapagos effect"—a phrase coined to describe the case of Japanese mobile phones in the 1990s. These phones, despite their technological advantages, never spread in popularity outside of Japan and became incompatible with foreign telecom networks; as a result, the Japanese word for such a phone is *garakē* (short for *garapagosu keitai*; or "Galapagos cell phone"). If competition were restricted due to the high barrier of entering such a complex marketplace, care robots might not be adopted internationally and would likewise become Galapagos technologies—existing and "evolving" in isolation in Japan alone.[10] On the other hand, enforcing proprietary systems could ensure the integrity of national technological intellectual property, protecting profitability from countries with cheaper labor and weaker intellectual property rights protection such as China. National authority in these matters depended on the working group in question. As Taylor explained,

> In the safety [working group], Japan has a lot more sway than the Chinese delegation because they've been there longer and they proposed most of the standards, they've done a lot of the legwork to actually develop the tests involved in it—they bought a whole [robot] testing center in Tsukuba to do all these tests, right?

Technical safety thus served as a source of international influence over the future development of the care robotics industry as well as a national competitive advantage and unique selling point for Japan's care robots.

Although METI and Japanese industry were enthusiastic about creating ISOs relating to personal care robots, the success of such a standard can only be judged once enough time has elapsed for it to be widely adopted, as can be seen in the case of ISO 13482. Between its creation in 2014 and the end of 2020, only around fifteen devices had been certified—all in Japan. One of the reasons for this is linked to another major output of the previous 2009–13 care robot project—the construction of a Robot Safety Testing Center in Tsukuba, mentioned by Taylor in the quote above. This was the only such center in the world at the time of writing and had been custom built to certify care robots to the ISO 13482 standard. Robotic devices could generally only be certified to the standard at this one testing center, and foreign firms had so far been uninterested in paying to have their products certified there. The testing center provided a physical symbol of the Japanese government's commitment to product safety and its leadership role in the international care robotics market. But if the standard continued to be ignored outside Japan—and indeed inside Japan, as the standard has not been integrated into regulations, meaning that certification is entirely voluntary—its adoption by only a handful of Japanese companies would call into question the ¥6 billion ($60 million) spent on the 2009–13 project as well as the international influence and relevance of the project. Indeed, an employee at a Japanese organization involved in advising companies on ISO 13482 certification explicitly described it to me as a "Galapagos standard" given the very low numbers of products being certified, noting that there had also been an ISO standard for Japan's Galapagos cell phones, but it had not led to them being adopted as an international design standard.

In theory, ISO standards constituted a focal point where consumer risk, corporate risk, and an engineering ethics of care robots primarily concerned with product safety could be harmonized. In part because alternative understandings of ethics are less readily—or perhaps not at all—translatable into standardized design features, mitigation of physical risk attains an equivalence with ethics, becoming something that appears quantifiable and thus controllable through standardization, as in the case of ISO 13482. In this way, the ISO becomes a kind of internationalized *ethical* standard, with Japan's leadership position in developing it signaling global leadership in the area of robot ethics, while transmuting moral questions to suit vested corporate interests.

"A Generic Human, Parameterized"

The standardization of ethical concerns about care robots in the form of globally scalable technical criteria of safety and risk management also required the standardization of human subjects. This can best be seen by considering in more detail how ISO 13482 was certified at the Robot Safety Testing Center. In June 2016, Matsumoto and the center manager gave me a tour of this cutting-edge facility, which is situated in a quiet part of Tsukuba and resembles a large warehouse from the outside, with colorful illustrations of people and different types of robots marching in unison on the exterior wall.

During my visit, the facility was empty and no robots were being tested. This may have been related to the time of year—we had not yet entered the "robot season." I was told that the facility was only really used for one or two months a year, based on funding and product development cycles at Japanese companies. At busy times, a careful protocol was in place to ensure that rival companies did not see each other's prototypes even when they were being tested on the same day. The process for a company to certify their care robot to ISO 13482 started with a risk assessment of the product, followed by relevant tests at the Robot Safety Testing Center. After a period of feedback and adjustment, the product would be certified as compliant to the ISO standard by the company itself, under the guidance of a certifying agency.

The center was divided into four main testing areas. The first room tested strength and durability, with machines set up to repeatedly exert different levels of pressure and other forces on robotic devices. A hangar-like second room was equipped with a motion capture system to record the movement of robotic devices. It also featured a running human dummy to test the speed at which devices could avoid a moving object, as well as a lamp that could mimic sunrays to test whether they interfered with sensors. At one side of the room was a large ramp to test devices such as walkers designed to automatically brake when going down a slope and drive themselves when going uphill, as well as different surfaces to mimic pavements and flooring found in care facilities and private homes. There was also a mock-up of the inside of a care home fitted with monitoring systems and other devices. The third room, designated for "interpersonal testing," contained an area to test crashing a robotic device into a child dummy. The final testing room contained a large futuristic-looking anechoic chamber with white paneling for electromagnetic compatibility testing.

For several tests, the center used American-made humanoid dummies based on standardized measurements of male American bodies, but a dummy designed according to the physical "specifications" of a standardized Japanese older adult had also recently been developed at AIST—a material representation of the abstracted physiology of an older care receiver used in technical analyses at RIRC. In this way, older Japanese bodies were themselves rendered as robotic analogues in order to open them up to further quantitative data collection, while disconnecting real end users even further from the development process. As in the case of Wabot House, a robot testing facility in Gifu Prefecture described by Jennifer Robertson (2018, 178–81), which was similarly built at great expense but rarely used after an initial media fanfare, the seasonal emptiness of the Robot Safety Testing Center further contributed to the sense that what took place there was primarily performative: the ritual certification of a decontextualized and abstracted safety and ethics based on robotic interactions with abstracted, indeed robotic, aged bodies.

Practices of "human informatics"—the title of one of the teams associated with RIRC—and the digital abstraction of older adults and the care they required took several forms in the work of the RIRC engineers. One researcher was collating data on the standardized bodily capabilities of older Japanese people to create a software model that could be used to design the specifications of robotic care devices. Another was researching the average activities undertaken by potential users of assistive robotic devices, using the World Health Organization's (WHO's) International Classification of Functioning, Disability, and Health (ICF). The ICF is a list of standardized activities broken down into individual physical movements such as lifting, standing, and walking; Matsumoto described it to me as "objective ways of describing the human functions." This breakdown of tasks into simple physical actions served as an ideal corollary for the breaking down of the life and care of older adults into standardized units, which in turn made them more amenable to processes of datafication and roboticization.

Likewise, a programmer on the team was working with a medical doctor on what he described as a "human body simulator . . . software that can simulate the interaction of the human body with external objects." He went on to explain this project:

> I think the ultimate goal is to create a generic human, parameterized. I think it's difficult to explain without having a software mindset, but you don't think—when solving such problems—you don't think about

a human, his bones, his joints, and general human body. You think in terms of activities—activities like walking. This is high-level abstraction, and you consider all your work is categorized based on activities—activities and the processing of the information . . . I have to use this specific word—it's like "interface segregation." Interface is a very well-defined contract between . . . if you claim you have a certain interface, it means that you should fulfill some contract, like walking. So an object with a walking interface can walk. And good design requires single responsibility. So it's obvious: when you can isolate complex objects and look only at this one interface—walking—you can design such a system where only walking ability is important, and you're considering only this thing. So the whole process—the success of the process—is when you can split responsibilities, not mix them. . . . This is closer to civil engineering and calculations like the types of simulations you need to conduct when you want to create a bridge.

I quote this team member at length because of the insight his description provides about the way in which human bodies and lifeworlds are viewed by many robotics specialists. Key to this application of software engineering principles to human life is the importance of splitting actions into individual discrete movements ("interfaces") as a starting point to breaking down physical interactions in care into discrete units for automation by robots.

As several scholars have observed, building models of robots expected to behave like, or interact with, humans also involves building models of humans and their capabilities. We can see this as a kind of applied anthropology—a particular mode of producing knowledge about and understanding humanness, human behavior, and sociality—that utilizes its own quantitative version of ethnography in the form of the abstracted, universalized, and fundamentally decontextualized analysis of measurements and data provided by databases such as the ICF and AIST's own Japanese Body Dimension Database. The jarring incongruity of the comparison of eldercare to bridge construction in the quote above highlights the level of abstraction and distance from the lived experience of those expected to use these robots which characterized engineering mindsets at RIRC. This way of understanding people and their lives largely detached from their social context and reduced to physical actions, movements, and interactions with the built environment is conducive to the idea of robotic replacement of human care labor. Moreover, in creating tools such as the human body simulator for private robotics companies to use in their product development process, researchers were actively encouraging them to utilize generic specifications specifically *without* the need to refer to or include real human users. I term this way of using abstract models to understand and design future care practices "algorithmic care" in the sense that an algorithm is "a precisely defined set of mathematical or logical operations for the performance of a particular task."[11] Algorithmic care represents an attempt to radically disentangle the messy elements of human life to produce a scientifically rationalized, Taylorized, and dehumanized version of care, materializing it in the form of robots designed to carry out each of the supposedly discrete and linear physical and communicatory tasks that are understood by engineers to constitute care.

Mathematical abstraction is not only a key element of robotics but also the basis of a status hierarchy among robotics engineers. Several interviewees explained to me that the roboticists seen as "most serious," who do "real" robotics, are those involved in the areas of dynamics and control, involving "higher" forms of abstraction. Dynamics deals with how the system of motors, actuators, and other components that make up a robotic assemblage work together to produce movement, while control refers to the programming of a robot in order to control it in motion. For a complex bipedal humanoid robot, it can take months to plan, program, and configure an action (e.g., walking) before the commands can actually be executed successfully. This is due to having to calculate, using highly complex mathematic models, the various forces at work and how they interact with each other, as well as how each individual motor will work together dynamically to create predictable movement in the physical world. Areas such as compliance—the flexibility of joints while walking—must also be taken into account. The dynamics of the robot must be built up in layers in a mathematical model involving algorithms translated into computer code. At an office party between members of a "serious" control team working with humanoid robots at the Intelligent Systems research institute and members of the RIRC team, the conversation turned to Romeo, a prototype humanoid robot developed by French company Aldebaran. Romeo was, they said, a more technically sophisticated robot than Aldebaran's other humanoid model, Pepper (discussed in chapter 6), but was abandoned when the company was taken over by SoftBank Robotics—an event lamented by several engineers. Although the members of the control team made fun of the "social" aspects of Romeo, including its questionable ability to guess the age and gender of its user by looking at them, they raved about its reversible actuators and other dynamic components.

One engineer told me that this elite top tier of roboticists tended to look down on areas of robotics such as robot vision and gaze tracking, which require a less complex understanding of mathematics and are essentially software-based, since they do not depend on a physical robot. At the bottom of this hierarchy is human-robot interaction (HRI), which is seen by many engineers as trivial soft science. I was told that HRI papers tended to be scheduled at the end of robotics conferences, indicating their lower perceived importance. In Japan, this status hierarchy has been somewhat complicated by a handful of high-profile engineers working on humanoid and animaloid robots, and several interviewees told me that there was some resentment in the international robotics community toward the direction taken by Ishiguro Hiroshi in developing cosmetically lifelike androids largely at the expense of dynamics or artificial intelligence, and indeed toward Shibata's Paro seal-shaped robot, since these robots made their inventors famous despite the fact that "they don't do much";[12] one researcher at RIRC dismissed most communication robots as merely "cell phones in a robot body."

This hierarchy based on abstraction summed up the way in which the social, communicative, and moral aspects of robotics seemed less important to many of the engineers involved in the Robot Care Project, both in the ways in which they thought about how the robots would be used and in their own professional lives. The imagined use of robots by end users seemed to parallel the conditions in which the RIRC team were working: with minimal human communication and with constant interaction with machines—primarily computers and sometimes robot prototypes—rather than people. This was also demonstrated when I asked interviewees whether they would buy the robots they were working on for their own parents. The question was met with some bemusement and laughter, with several team members telling me that they weren't sure whether their parents would like them but that they themselves would certainly enjoy playing with the robots. As Watanabe, a longtime RIRC researcher, put it, "For myself, I want them! They wouldn't help my lifestyle, but they're really interesting—I want them as an expensive toy! [laughs] But my own parents ... hmm ... not really.... Probably they wouldn't find them interesting."

Contracts, interfaces, bridge construction—these concepts seemed far removed from the daily lives of older adults. Lives and bodies are not abstract or universally regular, nor do they perfectly conform to average WHO measurements. Modeling based on standardized and abstracted older bodies runs the risk of failing to adequately represent the reality or wishes of older people themselves and their caregivers, who were rarely taken into account during the development process yet were the expected end users of these technologies. But if that was the case, then how were these robots, cast across the apparent gulf between research institute and care home, being received and adopted? How is robot care realized, appropriated, and used in everyday life? To address these questions, we now turn to Sakura public nursing care home in Kanagawa, which trialed three types of personal care robots being implemented through the Robot Care Project.

3

PORTRAIT OF A CARE HOME

Before looking at how robots were introduced into Sakura, we first need to consider what the care of older adults in a typical nursing care home in Japan involves. Who does care work, why, and how? What constitutes good care? It is essential to address these questions in order to grasp the impact of the introduction of robots and other technologies. The idea of care on which robot development at the Robot Innovation Research Center (RIRC) was premised was abstracted, deconstructed, and socially decontextualized. Actual practices and meanings of care are often glossed over by engineers and the caring dimension of social relationships assumed, because they tend to be seen as a "natural" and unchanging element of (female) human identity and sociality. In part because of the further assumption of a straightforward instrumental relationship between what robots are designed to do and how they are actually used, caregivers tend to be seen as irrelevant and are frequently left out of academic and popular discourses and policy debates about the development and introduction of such devices, particularly in Japan. The practices and meanings of care, however, cannot be taken for granted, and caregivers are crucial—though often overlooked—users of these technologies, who largely determine the success or failure of their implementation. In this chapter, we enter the genba (actual site) of Sakura to approach how the reality of residential care was happening in late-2010s Japan.

Sakura is a publicly funded nursing care home (*tokubetsuyōgorōjinhōmu*), the most common type of eldercare facility in Japan. During my fieldwork, there were around eighty residents, whose average age was eighty-nine, and average care level 3.9 out of a maximum of five, meaning that almost all had some degree of dementia or significant physical disability; more than half used wheelchairs. The home was situated in a quiet residential neighborhood, perched on top of a hill surrounded by greenhouses and small fields, although it had no garden and residents rarely went outside. The most able were taken for trips to local sights and restaurants by staff a couple of times a year, and some residents visited by their families were taken out for day trips.

Fifteen years old at the time of my research, Sakura was relatively new compared with the three other nursing care homes in the local area, but members of staff often described it as old fashioned because of its spatial layout—newer care facilities were often built as smaller units. Both the second and third floors used for residential care shared the same layout, with a long corridor separating a mixture of shared and private bedrooms on either side.[1] Each floor housed around forty residents. The second floor also contained a communal bathing area, where residents from both floors were brought for a bath on alternating days. At the end of the corridor was a large communal dining hall with a television on the wall that was left on most of the time. The dining area had several long tables where residents sat in chairs or wheelchairs, and this was where all group activities happened, serving as the focal point of daily life at the home, although residents often dispersed to their own rooms after mealtimes. Next to the dining hall was the workers' office, a cluttered room full of piles of old reports, bookshelves, posters showing exercises for care staff, and shift schedules. On the wall at the back of the office was a large panel of lights and bells that flashed or sounded alarms when calls were put through from residents' intercoms in their rooms or when certain residents stepped on a pressure pad in their room, indicating that they had stepped or fallen out of bed. The home was brightly lit and very clean, with light-colored flooring and walls decorated with neatly arranged colorful homemade displays of photographs and other materials, often relating to seasonal, natural, or national cultural events, featuring smiling residents enjoying special occasions or food.

Care staff wore colorful T-shirts, tracksuit bottoms, and sneakers, which distinguished them from the nurses and doctors who wore clinical white uniforms and suggested a less serious or professionalized yet friendlier image. There was relatively little day-to-day contact between staff stationed on different floors, and care workers generally worked on only one floor. Nurses, on the other hand, worked across floors, suggesting that it was seen as more important for care staff to maintain daily continuity in their care of residents and form long-term personal relationships with them. Care workers' official duties included serving meals, manually lifting residents—for example, from bed to wheelchair—taking them to the toilet, helping them wash and brush their teeth, keeping an eye on them (this is termed *mimamori* in Japanese), chatting, bringing them to and from their rooms, making beds, and so on. These duties were rotated regularly, and all permanent staff also took turns to be the day shift leader, which involved running shift handover meetings, updating electronic and paper records, and running the afternoon recreation session for residents. Although these responsibilities may seem neatly segregated, in messy everyday practice duties overlapped and ran into each other or occurred simultaneously, and the affective dimension of the work permeated and extended beyond individual physical tasks.

Care staff were around 60 percent female and 40 percent male—a higher percentage of men than the average for an eldercare institution in Japan.[2] Residents, on the other hand, were around 80 percent women to 20 percent men, partly reflecting longer female life expectancies and partly the fact that women often take care of and subsequently outlive their husbands and therefore are disproportionately more likely to end up in care homes. Although more men are becoming caregivers in Japan, the balance at Sakura reflected the still predominantly female-gendered norm of caregiving roles in Japanese society.

Managing Sakura and Introducing Robots

Mr. K had managed Sakura for the past ten years. He was a man of fifty who had spent a substantial part of his life in the United States and had a big personality and outspoken, pragmatic, sometimes cynical, views. In 2017, he was paying care staff around ¥1,000 ($10) per hour—a little higher than the local minimum wage.[3] He told me that "care work isn't *poorly* paid" because there was job security and additional government wage subsidies and bonuses available for employees. In addition, he said that care workers at Sakura were not subject to the same demands for compulsory overtime made on employees on similar wages in other industries such as hospitality or retail—a standard practice in Japan.

Given the labor shortage in the care sector and difficulty finding and recruiting new care workers, maintaining staff morale was one of his most important tasks, and he hoped that robots might help. He explained, "The nursing care staff are always looking for something new—they get less stimulation from the residents," suggesting the importance of the relationship between care workers and residents to care workers themselves and inverting the common perspective of caring as unidirectional. Mr. K also emphasized that staff were largely autonomous and made their own decisions about things. He said that he saw the Sakura management hierarchy as an inverted pyramid—with himself at the bottom—and surprised me by saying that he could not simply tell his staff what to do because they could just refuse. They apparently possessed this power because of their willingness to walk off the job at any time if it did not suit them combined with the ease of finding work at another care facility. This complicated the introduction of robots or in fact any change: staff had to be persuaded to accept them or at least to try them out. Mr. K said that staff tended to be highly skeptical of anything new and warned me before the start of the fieldwork that our plan to trial robots there would face resistance: "They're gonna say, 'Oh, we're too busy, we have no time, we don't have enough staff'—it's *always*—for *ten years*! It's the same shit. And maybe we are a little short staffed at this time, yes. But so is just about every other facility on the planet—well, in this country." He said that the best strategy would be to let staff see the effectiveness of using robots and how they could help both workers and residents and that he hoped this would win them over. The situation at Sakura was far from the instrumental management relationship assumed by many human-robot interaction studies in which robots are inserted into care homes seemingly unproblematically, usually for a few days or weeks, and used by staff perfectly according to the rules and conditions set by researchers. In fact, Mr. K's laissez-faire attitude to management seemed closer to that of the National Institute of Advanced Industrial Science and Technology (AIST) where, likewise, managers seemed unwilling to give direct instructions.

Recruitment of residents was also becoming increasingly difficult. The question of who would enter Sakura was a complex one. On the one hand, the law had changed in 2015 so that only people who had a care level of three or more were eligible to enter, and the home received higher subsidies from the government if they crossed certain threshold percentages of total residents with higher care levels. On the other hand, residents with higher care levels required more intensive care and had a shorter life expectancy, leading to a faster turnover rate and the need for continuous recruitment of new residents. Recruiting residents who required more demanding physical care and might have less ability to communicate also placed greater strain on care staff, which in turn could reduce morale and lead to higher staff turnover. An influx of severely impaired residents who might be expected to live for only a few months would in turn change the nature of the job the care workers were doing. Despite the relatively high morale among staff, as we will see, much of it seemed to stem from being able to communicate with residents and form mutually meaningful bonds with them over months and years.

Residents had to be found, assessed, and moved into the home quickly in order to avoid vacancies since the government subsidy for the home was also based on the overall occupancy rate. Sakura's care manager and counsellor both told me that although the home had a waiting list of around two hundred people, many of them did not meet the Goldilocks criteria of requiring just the right amount of care: without a serious medical disorder, without dementia so severe that it would make communal living impossible, but with a care level high enough to maximize the state subsidy. Family members trying to place a relative at

Sakura would routinely misrepresent their physical or mental condition on paper in hopes that they would be allowed to enter, and therefore an individual face-to-face assessment of each prospective resident was required—a time-consuming process. Many of the people on the waiting list were also on the waiting lists of many other care homes and might therefore drop out of Sakura's admission process at any time right up to the final stages before entry.

The question of who entered Sakura thus involved a dynamic balance between the care home as an economic ecosystem reliant on government subsidies and as a care ecosystem. In 2016, a certain type of person was being admitted into Sakura: mainly female, mostly from the generational cohort born in the 1930s, often with moderate dementia but somewhat mobile and not requiring intensive medical treatment. This was relevant to the question of which technologies could be used effectively: new residents were more likely to be reliant on pressure pads near beds, mobile toilets, walkers, and wheelchairs but less reliant on other technologies for bedridden residents (e.g., devices that could enable bedridden users to go to the toilet without the need for diapers) or medical technologies such as dialysis machines.

Attempting to introduce high-tech robots into this environment with staff who were perceived by Mr. K as resistant to change may have seemed something of a gamble. Nevertheless, the robot trial fitted into Mr. K's agenda of trying to keep Sakura at the cutting edge of the latest caregiving techniques and technologies. He regularly tried out new devices and invited companies to conduct demonstrations. Some of these technologies he adopted, including iPads to facilitate keeping and sharing digital patient records, and a couple of Roomba robotic vacuum cleaners. He said he made these decisions on a cost-benefit basis—for example, arguing that using Roombas meant less overtime for staff and improved job satisfaction since nobody enjoyed vacuuming, and staff did not rely on the additional overtime wages. As he explained, "The decision about [vacuuming] robots is mainly about cost, and it's not like anyone is losing their job to robots—yet."

He expressed an evolutionary view of technologies at Sakura, seeing value in the incremental introduction of different devices over time. It was therefore difficult to ascertain whether the robot trial was a serious attempt to introduce robotic devices in a manner that was likely to be expanded or a way to temporarily create the impression that further change was coming at some point in the future. He explained his initial rationale as testing the robots out to decide whether to purchase them in the future, exploring their potential for lowering staff costs, reducing accidents, and contributing to research. It later became clear, however, that he had decided not to buy or lease any of them. He emphasized that the trial was a "stepping-stone" and talked about how he was waiting for "real" subsidies from the government and for prices to come down significantly before he would actually consider buying them.[4] In particular, he described the robot Pepper as "not the answer but the first step"—part of an evolution—toward some better and cheaper robotic solution in a few years' time.

The project also seemed in part about continuing to project an image of innovation for the care home and trying to attract media attention. Mr. K was aware of the publicity potential of robots and seemed actively to court it as a business strategy. In his office, he kept a file of newspaper articles featuring Sakura, which he showed me on a couple of occasions. At the end of April 2017, he told me that he had contacted several local and eldercare industry newspapers and magazines to report on our research, to contribute further to Sakura's publicity file. During meetings with journalists over several days in May, Mr. K described the research as a form of "preparing hearts and minds" for the future introduction of robots, in order to "get the floor staff and residents used to a robotic environment."

At times, Mr. K appeared stressed and tired of his job, bored and frustrated with the day-to-day running of the care home—in fact, he often said as much. He spent a lot of time away in Tokyo attending exhibitions about the future of care and told me more than once that he was considering quitting to "try something new. But perhaps robots are the new thing that'll keep me interested." He also mentioned a couple of times that with ongoing consolidation in the industry, he was weighing whether to sell Sakura to a larger care home chain—raising the possibility that the project was also intended to make the care home more attractive for potential purchasers by garnering publicity in industry newspapers.

For Mr. K, as was likely the case for many other care home managers in Japan, robots served as a performance of innovation for several audiences: for current and prospective residents and their families; for care staff to maintain interest in their work and a generic sense of technological progress to support them in the workplace; and for prospective investors via industry publicity generated by the robots. They were a way to orient Sakura toward a future that was close but not quite there yet.

Care Routines and the Flow of Daily Life

In order to address the specificities of care at Sakura, in this section I present a composite narrative of what care staff and managers frequently referred to as the nagare or "flow" of daily life, based on two months of observation before the three care robots were introduced. The routine was largely identical day to day and replicated on both floors.

Residents started to wake up at around 6:00 a.m., and the following hour was an extremely busy time for the two night-shift staff on each floor, as they had to wake up the residents, help some of them to change their clothes and get into their wheelchairs, take them to the toilet if necessary, and then bring them into the dining hall. By 7:00 a.m., things were calmer—everyone was in the dining hall by now, ready for breakfast, and there was a handover meeting between the night staff and the incoming early shift staff. This meeting, like other shift handover meetings throughout the day, involved discussing each resident in turn, highlighting any changes or notable events over the course of the shift, including details such as how much they ate, whether they took their medicine, whether they went to the toilet, and their general condition. As the new shift arrived and the previous shift left, care workers greeted each other with a friendly *otsukaresama desu* (akin to "keep up the good work").

A near-identical scenario played out before each meal. First, staff would start to assemble residents in the dining hall, lifting those in bed into wheelchairs and wheeling them to the hall, or encouraging those who could walk or wheel themselves. After they had served tea to all the residents, a member of staff would make an announcement, telling residents what day it was, talking about the season or weather, and presenting some puzzles, riddles or word games, or memory exercises. This was followed by a few minutes of physical and speech exercises, and then the member of staff would choose a resident to call out the menu for the meal, which was written on the whiteboard at the front. After the resident called out each item on the menu, the staff member and other residents repeated each item in a chant. Different members of staff had different styles of doing the announcements and would often share jokes with the residents. Chiba, a good-humored, thickset man in his early thirties with a shaved head and mustache, would talk about each item on the menu, animatedly relating them to his own likes, dislikes, and personal experiences—an approach so successful that he often had residents gasping, "I want to eat it!" Meanwhile, other members of staff would bring out meal trays and place them in front of residents. All meals were prepared in a canteen on the ground floor, supervised by a nutritionist, and medicines were prepared by the nurses and administered to individual residents. Meals were always served simultaneously to all residents—eating together was an integral part of communal institutional life, bringing everyone on the floor together in time and space.

All rooms and toilets were equipped with call buttons, and some rooms with residents who were likely to get out of bed and wander around at night had pressure pads fitted on the floor next to their beds. These devices could trigger an array of flashing lights and alarm tunes from the panel in the workers' office. The alarms would grow louder if ignored, as frequently happened during busy times, leading to a constant background crescendo of electronic bleeping and telephones ringing. Residents could use an intercom system to call staff in the office from their beds, and likewise care workers would often use it to phone residents to ask them to come to the dining hall for a mealtime or activity. During hectic times—typically, before and after meals—the floor became noisy, with staff calling to each other and running backward and forward along the corridor.

After breakfast, the care staff were busy clearing plates and then helping residents back to their own rooms and to the toilet. For the next couple of hours, the floors were fairly quiet, as residents relaxed in the hall or returned to their rooms to rest, watch television, or read. Staff played music in the dining hall, and the mood seemed to lighten as residents became more active. This time of day saw the maximum number of care workers on hand as all four of the shifts briefly overlapped, so staff had more free time to joke around with and talk to or help residents individually. The contrast between the hectic activity of the time immediately after breakfast and the lull that followed was indicative of the way in which busyness was unevenly structured and activities distributed throughout the day. At around 11:00 a.m., staff started to bring residents back to the dining hall to prepare them for lunch. An experienced female care worker, Fujimoto, who has just been hired and is on a weeklong probation period, is asked by the floor manager to chat with residents for thirty minutes. Another care worker instructs her about one resident: "He's a bit hard of hearing, so talk to him from the left side." Fujimoto introduces herself to this resident using formal language and quickly starts a conversation.

A little later, a female care worker, Kubo, and male care worker, Otsuka, joke with an older woman with dementia, Inoue. They impersonate each other: Kubo introduces herself as Otsuka and vice versa. Inoue says that she wants to work. The staff say, "You're lying! Will you do anything?" Inoue replies earnestly, "Yes, I'll do anything." They say, "You can't!" But later, they sit Inoue next to the shift leader, Iwasaki, at the main desk in the office and give her some light tasks to do, referring to her as their intern. The tone of the joking is teasing rather than malicious, indicating a playfulness based on imagination and deception that I saw repeated frequently throughout the period of my fieldwork. Diego,[5] a Japanese Peruvian care worker with a deadpan sarcasm, sometimes pretends to residents that he neither speaks nor understands Japanese. Residents often seem to get the joke and laugh with him but occasionally become the butt of the joke if they fail to catch on that he is, in fact, fluent.

It was no coincidence that Inoue was often the target of (usually) affectionate teasing and attention. Inoue was a short-term resident who tended to make repeated requests of staff and according to several care workers had not yet "settled" into life at Sakura. She was described both as "cute" (*kawaii*) and as a "poor thing" (*kawaisō*), and her propensity to say and do inappropriate and unexpected things singled her out for special attention from care workers. Shortly after I started my fieldwork, care staff were in the office during a shift handover meeting, when Inoue quietly shuffled in without anyone noticing, inserted herself into the huddle, and interrupted to solemnly announce, "Today's lunch was delicious!" Everyone laughed. "She really likes her food, doesn't she?" "Yes, she looks so happy when she's eating!"

At other times, care workers employed imaginative or playful strategies to actively ignore residents who constantly asked for something—a counterpoint to active listening practices that will be discussed later. On one occasion, Fujita, an extremely cheerful and energetic care worker in her forties with short, cropped hair and glasses, replies to Kimura, an older woman who repetitively calls to her, "I can't hear you! It's no good!" This kind of response seemed a way for busy care workers to acknowledge residents and show that they were listened to, without having to take an action or being drawn into the sometimes-repetitive actions or words of those with dementia. At another time, an older woman gets up from her table and walks around to every resident in turn, politely saying, "Goodbye! You took good care of me. See you tomorrow!" A male care worker, Ikeda, calls to her casually, "Take care!" After she walks off down the corridor, he tells me drily, "She isn't going home. She always gets to wanting to go home around this time—always does the same thing."

Many residents were fairly quiet, but this did not necessarily equate to a lack of awareness of or engagement in the social life of the home, and several care workers explained to me that sensitivity in manners and language was extremely important to residents. On my first day of fieldwork, as tea was being served, I sat next to a quiet older woman. A care worker had told me that she, like many residents, was "shy." But when the care worker giving out tea did not hand one to me, the woman objected, "What about our guest?" When he obliged and served me tea as well, another older lady, who had been sitting silently on the other side of the table, smiled at me and said, "Be careful, it's hot!" Other residents were less communicative. One older male resident, Takahashi, who care staff told me used to be a famous jazz guitarist, wore a beanie hat and glasses and often played with a large teddy bear—members of staff referred to him as "a grandpa who likes teddies." He would talk to it quietly, chuckling, and then set it in front of him, resting his hands on it. Like many of the quieter residents, he rarely talked to other people and usually just smiled and nodded.

Lunch followed the same procedure as breakfast, and after the rush died down, staff again had a little more time to spend relaxing and talking with residents, and there was also more chatting among residents themselves in the dining hall or in their rooms. The CD player was again turned on to play Japanese pop music from the 1970s and 1980s or lively jazz. Takahashi is asked to help a care worker fold small towels, and she thanks him and tells him what to do. A couple of other residents join in, although Kimura forgets to fold the towels twice and produces a pile folded only once. Another staff member laughs at the mistake and asks her to fold them all one more time.

Recreational activities began at around 2:00 p.m. The choice of activities depended on the shift leader and reflected their personal preferences, with some choosing to tell funny stories or engage in banter with residents. Most shift leaders improvised or did the same type of recreation each time, which resulted in variation in activities on a daily basis as the role of shift leader was rotated. Some presented puzzles or games such as *shiritori*, a word-chain game where each player must think of a word beginning with the last syllable of the word given by the previous player. Others, such as Matsuo and Chiba, had a chat or discussion with residents, asking about their favorite foods or sharing an interesting story or funny anecdote. When Iwasaki and Chiba were on shift together—two male care workers and close friends in their early thirties who were both extroverted and funny—the banter took on the air of a stand-up routine reminiscent of *manzai*, a popular form of comedy in Japan involving two men trading jokes. Residents seemed to appreciate these engaging activities, chiming in and laughing along, although the entertainment was occasionally interrupted by residents asking for a drink or to be taken to the bathroom. In these cases, the leader would either have to pause what they were doing and help out or try to call over another care worker to attend to the resident.

Shift leaders who did not enjoy standing up at the front usually opted for showing a movie or setting up karaoke for residents to sing. Old classics, songs from school days, and modern *enka*[6] pop ballads were extremely popular among residents, who would sometimes sing solo or in chorus (sometimes with staff members), while others clapped along. The songs were often about home and nostalgia for past times and lost relationships. The most popular choice was *"Kawa no nagare no yō ni"* ("Like the flow of the river," famously sung by Misora Hibari in 1989, who died shortly after its release), a song about the ebbs and flows of life, which was played so often that it seemed a permanent feature of the soundscape of Sakura.

Recreation lasted between one and two hours depending on the type of activity, and afternoon tea was served midway through. As a young male staff member hands a piece of cake to a resident, he jokes, "That'll be ¥500 [$5]!" Other activities also often took place at this time, including regular visits from *keichō* (active listening) and choir group volunteers, a weekly music therapy session, and personal visits from relatives.

As the day wore on, several residents with dementia would start becoming restless and saying that they wanted to go home. The same pattern would be repeated numerous times each day: a resident in a chair or wheelchair would try to stand up and disentangle themself from the chair. A member of staff would rush over and ask the resident to sit down, explaining that they would be staying at Sakura that night. A couple of minutes later, the resident would try to get up again, and the pattern would repeat itself—sometimes for as long as an hour or two. These residents required a substantial amount of time and attention, occasionally leading to frustration on both sides. In one case, a staff member took a resident to her room, let her pack a bag, and even gave her a release form to sign to leave the home, although this was all an elaborate act—there was no possibility that she would actually leave.

After dinner, staff helped residents go to the toilet, clean their teeth, and get into bed. The nightshift leader went around the bedrooms helping residents change their clothes, dispensing sleeping pills to some residents, and measuring and recording vital signs. Otsuka, the shift leader one night, told me that residents were most likely to become disoriented and violent with staff at night while they gave out medicine, so a robot that could calm these residents down would be very helpful. Overnight, the two nightshift workers checked on residents regularly and responded to calls. Staff also kept an eye out for residents with dementia who would get out of bed during the night and wander around, sometimes crawling on all fours. One care worker told me that care staff often had to rely on their ears to detect any unusual noises coming from a room that might suggest a problem—he thought an electronic monitoring system would be helpful.

As Otsuka makes his rounds, he finds one lady sitting up in bed. She tells him she is hungry, and he goes to bring her some jelly from the fridge. He apologizes for it being cold, but she eats it happily. "It's delicious!" In another room, he talks at length to a resident who is asking for something but cannot articulate clearly what he wants. While this is happening, the alarm in the office starts to sound, indicating that someone is trying to get out of bed and stepping on a pressure pad, and another buzzer goes off, meaning that someone has pressed the nurse call button. Otsuka runs between rooms to check on each resident and ask them to get back into bed. He later explains that the most difficult part of the job is when residents with dementia repeatedly press the nurse call button—sometimes for a specific reason, such as to be taken to the toilet or because they feel lonely—but usually for no reason. Sometimes they press it because they want to go home and become violent when he goes to see them. In these cases, he says, distance is the best remedy: you simply have to separate yourself from them and let them calm down—you can talk to them, but they can keep talking and talking, and eventually he has to leave to check on other residents. He characterized this as the "tough" part of the job and termed the running around between calls, taking residents to the toilet, changing diapers, dispensing medicine, and checking each room as "this repetition."

At around 5:00 a.m., the night-shift workers eat breakfast before the floor becomes too busy again; at around 5:30 a.m., the first resident wakes up and walks to the dining hall to read a newspaper. Another comes to the office to ask for Yakult (a yogurt drink), and then goes back to his room. The corridor lights are switched back on, and care workers start to go from room to room waking up residents, greeting each with an energetic "Good morning!" Iwasaki wakes Saito by rubbing her cheeks, telling me, "Isn't she a really cute grandma!" The staff lift each resident who requires transfer into their wheelchair and then move them into the corridor. Then they push several residents in wheelchairs together into the dining hall to prepare for breakfast once again.

Who Cares and Why?

Thirty-seven care workers were employed at Sakura; their average age was forty-four.[7] Three were not Japanese citizens, including one Japanese Peruvian male care worker and two female care workers from Vietnam and the Philippines; all three were already resident in Japan at the time they started working at Sakura and did not enter via a migrant work visa scheme. Most of the staff had grown up locally and could be categorized roughly into three groups. The first were older male care workers who had almost all come from previous blue-collar jobs in manufacturing and warehouse logistics and had moved into care due to problems with their previous employer, sometimes in combination with personal events such as a family member requiring care. A second group comprised women from previous noncare, service industry jobs, or housewives, who worked part time and benefited from a long-standing tax exemption for those earning ¥1 million ($10,000) or less per year. Younger care workers mainly in their twenties and early thirties, a roughly equal mix of men and women, constituted a smaller third group. They had typically volunteered at care homes or facilities for people with disabilities during high school and gone on to study care at a vocational school. Due to the particularly acute shortage of young, trained care workers entering the industry, they had had their pick of institutions to choose from when they graduated. Overall, there was a strong sense that a care job—even if the pay was low—was nevertheless easy to get into and reliable employment.

Most of the staff said they had entered the industry because they wanted to care for an older or disabled family member or expected to do so in the future. Almost without exception, however, by the time they had completed their training or started a care job, circumstances prevented them from being able to care for this family member. For several care workers, this resulted in a sense of regret and a determination to care for others. For example, Iwasaki described how "my father had an injury in mid-life and—well, he's already passed away but—I was young and I wasn't able to care for him or help him, and I had some regrets about that, so I had an interest in this industry and decided to try . . . because of course I couldn't be filial, I was interested—[I thought,] 'Let me give it a try.'" Otsuka, a care worker in his thirties who had been working at Sakura for seven years since changing careers, entered the industry after his father developed dementia and died in hospital. He described how "I couldn't do anything at all for my father. . . . I was thinking all sorts of things and—well, it's not compensating for that or substituting for him, but—I thought I'd like to try doing a job caring for elderly people."

Starting a care job for the first time was jarring for some. Maeda, who had previously worked at a factory making engines for a large auto manufacturer, described it as "a world I'd never seen before—I was really disoriented. Having contact with [service] users with dementia, making conversation with the kind of people who until then I hadn't ever come into contact with, having to communicate—at first, I was really bewildered." Several care workers talked of a "culture shock"; Fujita described the gap between her expectation and reality:

> At first, I definitely had the hope of helping the cool grannies and grandpas—that kind of feeling—I came with a beautiful feeling. But when I really went [to work at a care home], there are all kinds of things, aren't there? For example, holding poop in their hands and then eating it, licking their false teeth, seeing those kinds of situations, I felt a little gap—bam!

Many had to contend with negative stereotypes of care work as dirty or depressing, especially among family members and friends. Nomura described how "initially, I had a very dark image of care—in the news there were stories of abuse, and I only heard negative things about it. I thought, 'It'll be pretty hard,' and I had the idea that I'd probably be bullied. But it wasn't like that [laughs]."

Others talked about the sense of responsibility involved in care. As Nomura explained, "I think the responsibility is greater than I thought. . . . Here you're involved in people's lives—it's a workplace where you're managing people's lives, so I think 'It's really heavy,' but that's actually the thing that makes it worth doing." Otsuka remarked on how the work of care is burdened with a responsibility that fails to be reflected in the level of pay and noted that "no-one does it for the money." Care staff said they found meaning in the responsibility and human connection inherent in their work. As Ikeda said, "I prefer a care job to my previous manual jobs because I find it rewarding seeing smiling faces. Then I feel glad I came. There are almost no [service] users who want to be here, but when I do something and they smile or they are feeling good about being here, the moments like these are rewarding and I feel fulfilled to be working."

Making Kin through Care

Despite this sense of responsibility, many care workers described Sakura as "a very easy place to work." This seemed at odds with usual descriptions of the Japanese eldercare industry, where the culturally prevalent image is that care work is a tough job. Pay was relatively low and most care workers suffered from some degree of back pain caused by lifting residents. Although they acknowledged these issues, staff at Sakura also said they had adequate breaks and holidays and almost no mandatory overtime for the majority of staff, which was unusual in Japan. As Mr. K explained to me, aside from actually raising salaries, great efforts were made to keep staff happy since they could so easily leave and find a different job at another care home. As he put it plainly, "Unless we maintain a really, really easy work environment, they'll just quit." Ishikawa described how "when residents enter other institutions, there's the idea that the best thing is for staff to ensure they have a good lifestyle. But the facility manager here was like, first it's most important to make an environment where it's easy for the staff to work. After that, they can perform well. . . . I thought, that's good, and that was the trigger [for deciding to work here]." This strategy seemed successful—according to Mr. K, the turnover rate among staff at Sakura was just 4 percent in 2015 and 7 percent in 2016, compared with the industry average of around 17 percent (Care Work Stabilization Center 2017).

Several care workers described the atmosphere as homely and relations with colleagues and residents as "like family." Fujita described how

> I think everyone is my companion [here], family. This is everyone's residence, isn't it—apart from the short-stay people. So I think it's a home. So I think everyone is family—mother and father, and also people I think of with a feeling of siblings—with that sense, I exchange conversation according to the person, and I think it's good if we have good relations.

Matsuo, a male care worker around fifty years of age, compared working at Sakura to "going to play at granny's house," and Fujiwara told me that "service users are like family—it's like my granny's here, and I have a grandson-like conversation. I care like a grandchild would." Despite the common official labeling of residents as "service users" (*riyōsha*), care staff frequently substituted kinship terms and occasionally also referred to residents as "elders" (*daisenpai*). This constituted something of a tension, however, particularly in the gendering of language use since politeness and formality toward "service users" were important characteristics of the Japanese service industry as well as a cultural ideal of feminine behavior in the postwar era in which the residents had grown up.

The common use of kinship terms by care workers and residents in addressing and referring to each other, as well as of informal language with those they felt closer to, contributed to what Diana Bethel has described as "an aura of intimacy" (1992, 113). Anthropologist Leng Leng Thang explained how such terms of "fictive kinship" or "vicarious kin" were not generally used by staff in the intergenerational Japanese care facility she studied because "residents are not blood-related to the staff" (2001, 84, quoting Lebra 1976, 88), although they were used when small children were brought in to visit. In the case of Sakura, the use of terms of address, including kinship terms, was complex and depended on individual care workers and their relationships with specific residents, serving to calibrate social distance. They could be endearing or infantilizing (in the case of the diminutive suffix—*chan*), or sarcastic (in the case of the playful suffix—*chama*—a combination of the honorific—*sama* with the diminutive—*chan*, so that *obāchama* would mean something like "darling granny"). New residents tended to be addressed in more formal terms, while those who had been living there for longer and with whom care workers were more familiar were more likely to be addressed with kinship terms.

This suggested one way in which the more transactional language of service provision was reconciled with ideals of filial care harking back to the ideal of the multigenerational ie family. Many residents seemed to acquire greater degrees of kinship over time, passing from being service users to substitute family members—becoming kin through the practice of care and gradual accumulation of a care relationship. Signe Howell describes "kinning" as "a process by which a fetus, new-born child, or any previously unconnected person, is brought into a significant and permanent relationship that is expressed in a kin idiom" (2003, 465).[8] Anthropologists have long argued that kinship relationships are socially constructed, even when they are also biological. "Kinning" is thus a more accurate way of describing this building of an intimate caring relationship than the phrase *fictive kin* which could suggest that such a relationship is premised on (self-)deception. It also aligned with the original intentions of many staff to care for a family member, with the ethical motivation of care based on kinship relationality that was transferred to some extent to other older people at Sakura in need of care. Family remained a key reference point in care work even in the institutional setting.[9]

Distance, Closeness, and Touch

When asked about what aspect of the job they liked the most, the vast majority of care staff answered that they enjoyed "communication" and "contact" with residents. This meant, as Onishi, a female care worker in her sixties, put it, "slowly communicating with users without being chased by duties." Similarly, Ishikawa explained, "Times when I can laugh with the service users are the most enjoyable." Care staff talked about their enjoyment in dealing with the unpredictability of certain residents as well as everyday banter. Iwasaki explained,

> Conversation with users—that's my favorite. With those with dementia, they say strange things that are total non sequiturs—absurd things. And even with trivial empty talk, they're communicating like family—[they] really like chatting about trivial things and I like talking. Just things like talking about that person's family, my own family, empty gossip—for instance, "I had a coffee at Starbucks; it was really tasty"—things like that—really empty chat.

Many care workers employed spatial metaphors to describe their relationships with residents. One told me, "I feel close to those who can communicate. We can understand what each other says, they can remember your face—it becomes a close relationship." Staff talked of "approaching" or "getting close to" residents in the sense of reducing social distance and gaining familiarity with them. Another worker explained how this could help users: "Quite a lot of users who say 'I'm lonely'—they're happy if they can just talk a bit. I think it's good if I can get close to them just a little."

Tactility was perhaps the most important way in which care workers "got close to" residents, both literally and, in turn, metaphorically. I often saw staff hugging, patting, rubbing, tickling, nuzzling, and massaging residents, occasionally putting their arms around those they were feeding at mealtimes, giving little touches as they passed by, and generally sharing a great deal of bodily contact, which care workers referred to as "skinship"—literally, the construction of kinship through touch. As one told me, "I have a visceral sense [literally "a feeling in my skin"] that generally [the residents] think of us as family." One day, I saw Chiba sit next to a female resident, smile at her, and say, "Hey, hey," take her hand and rub her palm while they watched television together—an almost-familial scene of domesticity. Care staff often held hands with residents to guide them along the corridor.

This level of physical contact calls for some explanation, given that in Japan physical touch between adults—even between close friends or relatives—tends to be avoided. Diana Tahhan writes in her 2014 book *The Japanese Family: Touch, Intimacy and Feeling* about the importance of skinship and touch between parents and young children, which usually ends abruptly at around the age of five. She argues that touch and emotional closeness continue to exist in the form of what she calls touching at depth—a nonphysical form of intimacy that lingers in the tangible "inhabited" space between people, especially family members, enabling a continuation of intimacy. Touch was perhaps particularly important at Sakura given the extensive use of face masks by care workers during flu season even before the COVID-19 pandemic and the fact that a considerable proportion of residents was hard of hearing. Yet beyond these reasons, eldercare at Sakura involved a return to the importance of skinship and intimate touch in old age—touch evoking a powerful sense of kinship—which care workers told me contributed to "peace of mind" (*anshin*) among care recipients.

Joking is often overlooked in ethnographies of institutional care, particularly in Japan, where the tone of such works has sometimes, and perhaps for understandable reasons, verged on the bleak. Jokes, however, worked in conjunction with both informal language and tactile care, putting residents at ease and helping to overcome embarrassment and negotiate potential discomfort caused by intimate touch or the vulnerability and dependence revealed, for example, during lifting and transfer. It served to mark and smooth out points of ambiguity and tension in daily care: for example, helping to transform the meaning of physical contact which, though intended to provide peace of mind, could otherwise potentially also signal unfamiliar, unwanted, or coldly clinical intimacy. Indeed, like the use of kinship terms, engaging in creative as-if play enabled both staff and residents imaginatively to transcend and, in a sense, escape the quotidian routine of the institution. The need for a joke to get a laugh enabled a further release from the institutionally determined roles of active care worker and passively dependent service user by affording a degree, however small, of reciprocity. During my time at Sakura, I observed care staff joking about themselves—for example, lamenting their losses at pachinko (a mechanical gambling game popular in Japan) to the gently disapproving yet amused tuts and chuckles of an audience of residents. They also joked about the residents, at times pretending they shared a romantic relationship, or occasionally addressing a resident in a tongue-in-cheek way as *sensei* (teacher) and asking for some words of wisdom. A joking, tactile relationship with degrees of kinship, combining the verbal and nonverbal, during individual time spent one-on-one between care worker and resident, was inseparable from other elements of care provided at the home.

It is important not to idealize such relationships. Making residents like family members and reducing institutional distance exposed the emotional vulnerability of care staff and raised the stakes involved in relationships. As Mr. K told me, "Staff have to deal with people dying that they really like, that they've taken care of for years, so it's like having your grandmother die on you repeatedly over and over and over and over again." Many care workers described sustaining minor injuries from residents who lashed out during transfers, particularly new short-stay residents, and as we will see, transfer itself contributed significantly to injuries. A few staff had favorites, and some residents received far more attention than others. Kinship feelings, though common, were not universal: two care workers, in contrast to the others, rejected the idea of a familial relationship with residents. These views were, however, in the minority.

Personalized Care and Building Trust Relationships

Care staff frequently used the phrase *human to human* to describe the character of their care work. As Onishi put it, "The other person is human, and residents are all different." For care workers, the word *human* (*ningen*) was often talked about in terms of personhood and individuality. Care workers were particularists, in contrast to the engineers at RIRC who dealt in abstract generalizations. A typical statement came, for example, from Iwasaki: "Everyone is an individual. You have to understand them and treat them as individuals with their own character." This was reflected in the ways in which care staff introduced residents, pointing each out separately, listing their likes and dislikes, and describing their personality.

It also figured in how care workers defined "good care," which involved what they called "individual" or "personal" care: a focus on individuals and their specific needs. Several also talked about how this represented a general change across the care industry. A greater emphasis on individuality seemed loosely to have coincided with the introduction in 2000 of the Long-Term Care Insurance (LTCI) system described in chapter 1, together with ongoing processes of individualization in Japanese society more broadly. In the context of the national discourse of muen shakai—"a society of no ties," where traditional relationships with family and friends have broken down—"personal care" seemed to mean two somewhat opposed things to care workers. On the one hand, it represented a shift toward treating residents as "service users" in a commodified transactional relationship; and on the other hand, it meant caring for residents as individual persons, often through the idiom of family and filial values. For care staff, these two aspects of "personal care" had to be harmonized with the constraints of the institution and its spatiotemporal flow of daily routine.

Fujita had worked in the field of care for sixteen years, including three years at two nursing care homes in the early 2000s followed by a twelve-year stint at a hospital before coming to Sakura. When asked how the previous care homes where she had worked differed from Sakura, she answered,

> This was over twelve years ago, so probably the times were quite different—I think how things are done has changed from how they were done at that time. But at my previous workplace, the users' clothes and even

underwear were all communal. So everyone was together, they didn't have names—not [treated as] individuals but all lumped together.... There was a tatami room next to the office like there is here, and basically the short-stay people would sleep there in a huddle.

This description tallies to some extent with Bethel's ethnographic account of Aotani Institution for the Elderly in Japan in the 1980s in which the underlying tension between individual personhood and institutional communal living played out in residents' spirit and practices of resistance to the set rules and conformity of the care home. Clearly times had changed, but such tensions had not disappeared; in the case of Sakura, however, it was the staff rather than the residents who were resisting the homogenizing power of the institution. Several care workers complained about the institutional nature of Sakura. One senior member of staff compared it to a previous smaller unit facility in which she had recently worked:

> The number of staff was I think the same in proportion to the number of users, but we were able to do individual care. Here no matter what, it's an institution—in a fixed institution, it adjusts users while in the previous [place I worked] we adjusted to the users. So if you ask me which has the higher quality of care, the previous one did. But one way or another, if it comes to an institution with this number of people, I think there's no way around that.

The quest for personalized care that treated residents as individuals, as humans, was pursued by all the care staff, and the most important tool for this was communication in order to get to know residents on a personal level. As indicated by the use of kinship terms and the importance of tactile care, personalized care tended to be equated with personal care—yet, this butted up against the seemingly fixed institutionalism of the care home.

I asked care workers to explain how they built trust relationships with the people under their care. In addition to the use of kinship terms and touch, Matsui answered, "First of all, when it comes to doing care, you have to *know* the users. From ADLs [activities of daily life] to what they like, what they dislike—based on these, you can steadily use the kind of topics where you can talk and have communication based on their own circumstances." It is notable that most of the chats between care staff and residents during recreation centered on sharing personal preferences and experiences. Gradually accreting knowledge about residents over time in exchanges that increased mutual familiarity also contributed to good care in other ways: Ota explained how such familiarity with residents' peculiarities enabled care staff to notice illnesses before they set in. Building relationships with individuals through interactions over time is, of course, an inherent part of human sociality. It is important, however, to contrast these human behaviors and relational skills with the instant and ahistorical digital relationships performed with robotic devices that do not currently, in a meaningful way, store memories of interactions that can be used to fine-tune future interactions and thus accrete dynamic, let alone empathic, relationships with individuals over time.

Care staff also directly linked care to time, relating care to the manner in which time was passed by residents. Many described good care in terms of trying to make "days easy to live[10] and enjoyable" for residents. Goto talked about creating an "environment where it is easy to live" and stated that "the best care is making sure residents can live their lives [to the full] one day at a time." Nomura observed how, on "days when users seem to really enjoy passing the time—then I think, 'I really did a good job today!'" Ease and comfort cut both ways, with the sense that they mutually benefited both worker and resident. Staff talked about the goal of doing the job "with ease" or making the job "easy to do" by making days at the home "easy to live" for residents. This understanding seemed to conceptualize the "burden" (*futan*) of care not as a zero-sum game where a weight was shifted from residents onto care workers but rather as a dynamic relationship that could be rebalanced, relaxed, and improved with time. Maeda described how

> good care is relaxed [*yutori*] care—for both staff and users. If either are busy—if the staff are busy, it's difficult for the users to make requests. Because I think probably it's not possible to grant their requests, so I think I want to increase the number of care workers on duty at busy times. If we do that, then the staff, in terms of their feeling, can do it slowly, and calm down, and can be kind to users who ask for something, and I think that if there's less impatience, that's linked to fewer accidents, too.

Time, Space, and Surplus Capacity

This central tension between relaxed care and busyness played out in the key interrelated elements of the institution: time and space. Care staff raised two main complaints about Sakura. The first was based on the built spatial environment of the home, and the second was about the availability and distribution of time. Several care workers told me that the building was not well laid out for care. The toilets were too narrow, the bath and changing rooms were too small, and the workers' office was placed at the end of the overly long corridor rather than in the middle, where the staff on duty could more easily keep an eye on residents. The placement of tables and chairs in the dining room made it difficult and therefore dangerous for both residents and care workers to walk around. The staff break room was too small and cluttered. The awkward distribution of physical space interacted with that of time—the two worked in tandem to produce the distinctive flow (nagare) of life at Sakura.

The largest concern, however, was about the institutional regulation of time as a resource. As Ishikawa put it,

> The way of doing care here is according to time—there are a bunch of set things to do and only a set number of staff, so you're busy with the feeling of how many minutes and how many seconds you have [laughs]. At the group home [where she had worked previously], there were nine residents so we had yoyū, we could spend time individually, talking with them, drinking tea together and so on—so being able to care while doing those things, that's a big difference.

Several care workers used the expressions "chased by time" or "chased by the job" to describe having to complete all their duties within the time available to them under the fixed institutional schedule. Otsuka explained,

> If you're like "next after you do this, you have to do that, then after you do that, you have to do this," then of course you absolutely can't respond to users' requests. Even when they ask, "Excuse me, can you do this?" you have to reply with something like, "Oh, sorry. I'm just doing something right now and I can't take my hands away from it. Let me do it a bit later"—it's no good, but somehow it ends up like that. And then if a user doesn't calm down, it takes all your time on just that one resident. It's a bit strange to say they take away all your time, but if you look at it from that point of view, the people around them, those other people's care ends up being neglected. So I want to do something about that situation. After all, if there aren't enough staff, you can't have care with yoyū.

The term *yoyū*, used by both care workers quoted above, refers to a surplus capacity: enough, with something left over. Yoyū was used to refer to the spatial or temporal capacity for care and was also sometimes employed by care staff metaphorically. For example, Otsuka described how caregivers need "capacity in their hearts [*kokoro ni yoyū*] to be able to do good care." The one term thus brings together both subjective and objective resources and arrangements of space and time for the specific end of care. As with the word *yutori* (leisure, relaxed, with time/space to spare), which was also often used by care workers, it represented an ethic of care unbounded by and resistant to the institutionalism of the care home, as well as the productivist ideology of wider Japanese capitalism.

At Sakura, the ethics of personalized care affirmed by most of the care staff, often based on their own experiences of wanting to care for family members, involved a holistic ideology of bodily, social, affective, and ethical care practices, hybridizing "traditional" ideals of filial caring relations rooted in the imagined multigenerational ie family with a newer emphasis on personalization from the LTCI system, which involved practices centered on treating residents as individuals. Care staff tried to harmonize this hybrid ethics of care with the institutional rigidity that governed the "flow" (nagare) of space and time across the fixed schedules and routines of institutional care at Sakura, through the idiom of kinship, and through concepts such as yoyū (surplus capacity of time/space) and yutori (leisure) that resisted capitalist logics of labor efficiency and service usership.

This was a stark contrast from the view of care among engineers at RIRC as an abstracted, linear series of discrete care tasks seemingly disconnected from human-human interaction or social relationships. And whereas communication was antithetical to the work culture of the engineers at RIRC, it was essential to the provision of care at Sakura. Changes to care policy at the national level that had altered the composition of residents entering the home, with new residents on average frailer and in need of more labor-intensive care, increased the importance of bodily care and communication through touch even further.

Yet, if care and the ethics of care are constituted in material practices and human relationships that are mediated through configurations of space and time of the care home, this also highlights the importance of the built environment and the care technologies that are embedded within it. It was into this care landscape that the robots were introduced. The next three chapters each focus on the introduction of a different care robot and the care activity that it was intended to reconfigure. Having heard some of the aspirations for care robots from Mr. K and his staff, we now turn to how robot care was put into action.

4

HUG

Reconfiguring Lifting[1]

At 6:00 a.m., Iwasaki, a gregarious male care worker in his early thirties, woke Yamamoto, who was lying down in her bed, by rubbing her arms and cheeks and telling a joke. He pulled her empty wheelchair over and positioned it next to the bed. He then put one arm around her neck and shoulder and the other under her knees and raised her into an upright sitting position at the edge of the bed. He remarked that aged bodies are stiff and bent but also very fragile, so he needed to shift Yamamoto to her side first as she could not straighten her body. He then asked her to hold on to him and, putting both arms around her, bent his knees, lifted her up, and swiveled to maneuver her into the wheelchair in one fluid motion. He removed her outer shirt and replaced it with a fresh cardigan and then put on her socks and slippers. Throughout their interaction, he chatted and joked—he claimed that residents loved his dark humor and crude jokes, and Yamamoto's chuckles suggested she was no exception.

Every day in care homes across Japan, care workers manually lift older adult care receivers. Lifting is a mundane activity that epitomizes the "burden" (*futan*) of care in contemporary Japanese society, placing its weight squarely on the backs of caregivers, and performing both the literal and metaphorical closeness and dependence of the care relationship. It represents the sharp end of the care profession, and endemic back pain, reported by a majority of care workers and directly attributed to lifting,[2] is one of the main reasons for the job's reputation as "3K" (*kitsui, kitanai, kiken*): difficult, dirty, and dangerous. As a result, back pain due to lifting is often identified in state and media discourses as a major contributing factor to the large and growing shortage of care workers in Japan.

Unlike several other higher income countries including the United Kingdom, Australia, and New Zealand, Japan lacks a nationwide safe patient handling policy[3] intended to prevent the need for care workers to lift care recipients manually by implementing mechanical lifting equipment and procedures for its use. Matsumoto, the Robot Innovation Research Center (RIRC) team leader, told me:

> It's strange that there is some regulation [that] we cannot lift more than thirty kilograms' [sixty-six pounds'] weight in the Japanese working environment, but it is not clear that humans are one of the objects that it applies to. Originally it was a regulation for laborers working at ports or in mining.... But it's obvious that most people are heavier than thirty kilograms even in Japan. But so far, the government doesn't really clearly say that it's in regulation or not.... Injury is not really regarded as an accident in the care workplace in Japan. Many people suffer from back injury. They complain but just ... [shrugs]

It is reflective of the low socioeconomic status of care that the physical suffering of care workers has long been disregarded—officially acknowledged but neglected in terms of legal protections. In Japan, as in other countries including the United States, care homes are among the most dangerous places to work.[4] In other countries, this toll might tend to fall on migrant workers; in Japan, it falls disproportionally on middle-aged Japanese women.

Some efforts have been made by Japanese health organizations to introduce occupational health guidelines related to manual lifting in care,[5] and some care home groups have introduced their own safe patient handling policies in an attempt to boost staff retention. But rather than implementing a national safe patient handling policy using existing relatively low-tech mechanical devices such as hoists, Japanese government action has instead focused on high-tech robots. One of the key deliverables of the government's 2015 New Robot Strategy was to reduce back injuries to zero through the use of robots, and this was also a key objective for the National Institute of Advanced Industrial Science and Technology (AIST) in implementing the Robot Care Project.

Robotic Lifting Devices

Lifting may seem the lowest hanging fruit for care robot engineers—a repetitive, clearly defined manual labor task that poses a substantial health risk to tens of thousands of care workers in Japan every day, thus representing an immediate need with quantifiable benefits, which involves a seemingly straightforward adaptation of industrial robotic technologies that already exist. In the Robot Care

Project, lifting devices were divided into two categories: wearable and nonwearable. Wearable devices include machines such as Cyberdyne's HAL (Hybrid Assistive Limb) exoskeleton, which can be worn by caregivers and uses electrical motors to augment the weight that a user can normally lift. Other types of power suit use compressed air or elastic material to transfer loads from the lower back to the knees and thighs. Nonwearable devices include machines that use robotic arms to lift users, such as Fuji Machine Manufacturing's Hug and Muscle Corporation's Robohelper Sasuke.

Yet these devices, far from being gratefully welcomed by beleaguered care staff as anticipated by government strategists and robotics engineers and researchers, have instead met with considerable, albeit tacit, resistance from care workers and care home managers. Despite the recent proliferation of the types of robotic lifting aids available (in addition to relatively low-tech and low-cost hoists that have been on the market for decades), they are far from commercially successful and have so far failed to make a significant impact in Japanese eldercare. In a 2018 interview with *The Guardian*, Hirukawa, the head of Intelligent Systems at AIST, stated that lifting robots were only implemented in about 8 percent of care homes in Japan, in part because of cost and in part because of "the mindset by the people on the frontline of caregiving that after all it must be human beings who provide this kind of care. . . . On the side of those who receive care, of course initially there will be psychological resistance" (Hurst 2018). In fact, even 8 percent was likely an overestimate: in a 2016 survey of almost nine thousand eldercare institutions, less than 1 percent reported the use of a lifting robot (Care Work Stabilization Center 2017); by 2019, this had only risen to 2.3 percent (Care Work Stabilization Center 2020).

Whereas other countries in Europe and North America have embraced lifting aids as integral to the implementation of safe patient handling policies, even Japanese care homes that have robotic lifting devices or manual hoists rarely use them. Matsumoto told me:

> In Japan, only 5 percent of nursing homes have hoists, and actually even if they have them, they don't really use them. People working in hospitals and nursing homes don't like technologies. [They think] manual treatment is better than [using] the machines. . . . So you have to show them how using the machines, the effect of using machines, how it is safe and reduces the risk of injury in most of the caregivers and patients—because otherwise, the new technologies like the robots the companies are developing cannot really be used in this area.

I heard similar views from those working in care. A care home manager in Ibaraki Prefecture told me that he did not consider purchasing a lifting device because "so far, institutions that have bought lifting machines haven't used them." In this chapter, I focus on the Hug lifting robot, examining staff resistance to its introduction in the context of the practices and ethics of care at Sakura. Sakura was typical of nursing care institutions in Japan in terms of the continued use of manual lifting, a relatively high rate of self-reported back pain among care workers, and a general rejection of mechanical lifting devices even when these were available for staff to use. There clearly seems to be a disconnect between the technoscientific understandings of care among robot engineers and government officials and those of staff at the genba (actual site) of care. But what is the nature of this disconnect? Why does there seem to be so much so-called psychological resistance from care workers to adopt these devices and save their backs?

Lifting and Tactile Care at Sakura

At Sakura, care superficially resembled a repetitive logistical operation—somewhat akin to one care worker's previous warehouse job that involved "moving things in the right order" (*buppinseiri*). Residents also had to be moved at the right time, to the right place, and in the right order, according to the institutional nagare or "flow" of daily life and care at the home. Throughout the day, staff constantly moved residents between their beds, toilets, dining hall, and bath, and these movements often involved lifting. Out of eighty residents across both floors at Sakura, forty-nine required manual transfer. According to data gathered by staff at the start of April 2017, counting every single instance of lifting across the two floors,[6] care workers performed more than four hundred lifts over a twenty-four-hour period. On average, each resident requiring transfer was lifted at least eight times a day. The majority of these lifts were concentrated in the morning and evening, and unsurprisingly, care workers frequently referred to these times as the most difficult of the day, particularly as they coincided with the lowest staffing levels.

The burden of transfer seemed to be reflected in the results of an initial survey of all thirty-seven care staff that I conducted in March 2017, prior to introducing Hug. Eighty-six percent of all staff reported some form of back pain, with an average pain intensity level of 4.2 on a scale of one to ten.[7] Although transfers were carried out using "body mechanics" techniques—standardized ways of lifting intended to minimize the risk of injury to both the member of staff and resident being lifted—several care workers stated during interviews that their back pain had increased over the years of working at Sakura. Staff who had been doing the job for a longer time and suffered from a greater level of back pain were more likely to have developed means of dealing with the burden of lifting, including using a lumbar support belt (a wide belt around the waist that supports the back muscles and spine) or using adhesive pain relief patches or painkillers. Three care workers had suffered herniated discs while working at Sakura, which they attributed to lifting, although according to Mr. K, during his ten years as manager of Sakura, no staff member had complained to him directly of significant back pain or injury, and no staff member who quit their job gave back pain as a reason for their departure: "I've never heard of any staff member complain that lifting is what makes the job unattractive."[8] When injuries occurred, 70 percent of the cost of medical treatment was covered by the state under the national health insurance system, with the remaining 30 percent paid by Mr. K, who would also pay the employee for the duration of their sick leave. Mr. K appeared unconcerned about the possibility of lawsuits, and it seemed rare that a care worker in Japan would sue their employer over such an injury.

These figures on back pain at Sakura broadly correspond to the results of larger studies looking at care staff working in eldercare institutions in Japan. In a survey involving 1,925 institutional care staff, 58 percent reported back pain, a figure that rose to 81 percent when including those who had experienced it in the past. In terms of pain intensity, 55 percent reported that they "sometimes feel a light pain" (Ueda et al. 2012). More broadly, in 2016, "catastrophic" back pain (defined as leading to more than four days off work) accounted for the vast majority of workplace injuries among workers in the health and hygiene industry nationwide, which includes care workers—1,423 out of 1,540 according to the Ministry of Health, Labour, and Welfare (MHLW 2016). It seems very likely that there were many more unreported injuries that failed to meet this threshold.

Otsuka, a care worker who had previously worked in a warehouse without injury, seemed to confirm Matsumoto's earlier point about back injuries not being regarded as a workplace accident in the care sector. He described the circumstances surrounding his own slipped disc at Sakura and was characteristic in playing it down:

> O: I had a herniated disc recently—in December last year. Sometimes my back pain gets pretty bad, but I thought it was strange because the back pain was a bit prolonged—so I went to the hospital and went to the orthopedic surgery and was told that I had a herniated disc [laughs]. And I rested for two weeks.
>
> JW: Was it because of transferring residents?
>
> O: Probably all sorts of things—I think the fatigue had built up. Of course, there were also transfers—I think there were all sorts of things.
>
> JW: What do you think about that?
>
> O: Care workers often have body pains—in their knees, back, and shoulders. If you mess those up, then you get so you can't do your job.

> When I got the slipped disc, I thought, "I have to quit" and "I have to leave this job." But well, now this herniated disc, if it's not so bad they don't need to operate on it—well, I didn't really know much about herniated discs, so I was thinking, "If you get it once, it won't heal; if you don't get surgery, it won't heal"—and if so, then "I can't do the job, right?" But it wasn't like that, and anyway in my case, it seems that it wasn't so bad.

Yet, tactile contact with residents also had many positive connotations at Sakura. Staff frequently talked of the importance of touch in building and maintaining a trusting relationship with residents, to become "like family." Several care workers told me this was particularly important when caring for residents with severe dementia or those who could not communicate verbally—touch became a technique for affective communication, regularly conveying and reperforming familial intimacy and building a "close" relationship through entangled physical and social care practices.

Lifting is a significant part of this tactile care, reflecting a complex bond between the vulnerable and dependent bodies of care recipients and those of caregivers. The moment of lifting and being lifted can reveal physical discomfort as well as fear of inflicting or receiving injury for both care worker *and* resident—a co-vulnerability, particularly if either one is not used to the maneuver or if the worker is unsure of their own strength and ability to lift. The exact form of lifting varied according to residents and their physical abilities and sometimes involved two staff members lifting together. Several care workers told me that there was a gap between how transfer was supposed to be done according to the textbook and how it was actually done in real life, partly because aged bodies come in many forms that often do not correspond to standardized textbook examples. As Matsui stated:

> At first, well, [I thought that] the users are all grandpas, grandmas, so I was a bit scared to touch them because they're weak. But as I was doing it, I started to get a sense of how much physical strength was OK to use. And there are some difficult things like contractures [a condition of persistent rigidity in the muscles or joints]. You need to think, "How can I safely deal with the burden?" Users' shapes are all different—you just have to build experience.

Care workers have to respond to verbal and nonverbal cues from residents in order to transfer comfortably and safely. They also have to take into account their knowledge of the personalities and characteristics of individual residents: for example, whether they might strike out while being lifted, as a small number of residents suffering from dementia sometimes did. In this way, transfer can be understood as a skilled bodily technique dependent on perception, touch, and empathy.

Introducing Hug

Hug is a mechanical lifting device measuring just over three feet tall and weighing 143 pounds, which was brought to market by Fuji Machine Manufacturing (hereafter, Fuji) in 2016. Like other nonwearable lifting robots, Hug was developed by an industrial robot company and was the result of commercial diversification: tweaking and repurposing an existing industrial robotic arm for the eldercare market. The name Hug seems to refer both to the way in which the user "hugs" the robot, putting their arms around the lifting pads and gripping the handle on the other side, and to how the machine in turn "hugs" the user back with its padded robot arm. The name implies an intimate, caring relationship between user and technology somewhat at odds with its industrial appearance, which even Fuji's managers admitted, in an interview I conducted, "scares 80–90 percent of elderly people at first." They told me that some of the benefits of its use were that heavier people could request a female caregiver to perform the transfer without feeling embarrassed and that older and even pregnant care workers could continue to do transfer by using Hug.

Hug is fairly straightforward to operate. The care worker first positions it toward the resident so that the arms of the robot extend under the armpits of the user and then gently pushes the user forward until they are leaning on Hug and gripping the handle at the front, with their knees resting on a knee guard at the base of the device (as shown in figure 2). A simple control pad is then used to operate the robot arm on Hug: first, the arm rotates and tips the user farther forward until most or all of their weight is resting on Hug, and then it lifts them up to a near-standing position. The resident's weight is supported by Hug and distributed across their armpits, their arms gripping the handle, and their legs, to the extent that they can take any weight on their legs. The care worker can then wheel Hug to position the resident over the wheelchair, bed, or toilet seat, and then go through the same operation in reverse to complete the transfer. The whole process takes about ninety seconds. In 2016, Fuji claimed that Hugs were in use in around fifty care homes, mostly located in Tokyo. At the time of my fieldwork, it could be purchased at a cost of ¥1.44 million ($14,500) spread over five years.

Mr. K first brought Hug to the attention of his staff through a demonstration and training session carried out by Fuji employees at Sakura in December 2016,

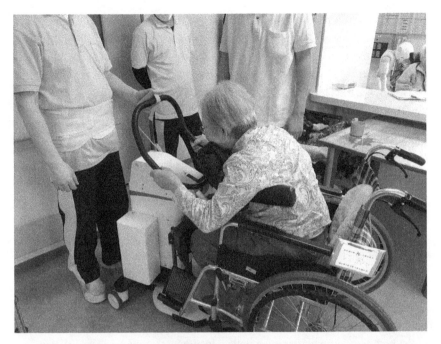

FIGURE 2. Hug in action. On the left is Otsuka, who is operating Hug. He is wearing a support belt after suffering a herniated disc, which he attributed in part to transferring residents. Pepper can be seen in the background. Photo by author.

which lasted about an hour and was attended by half the home's care staff. A survey of the attendees was carried out immediately after the session, and the feedback was strikingly negative: while 60 percent of staff members involved in the trial thought that Hug was easy to operate, and 80 percent thought that it could help reduce the burden on their backs, only 15 percent thought it could be used with peace of mind (*anshin*) by residents, and a mere 5 percent thought that many residents could use Hug. Out of those who provided a response to the question "Do you want to use Hug for transfer?" almost half said that they did not want to use Hug; only a quarter said they wanted to use it.

A later survey of all care staff at Sakura, which I conducted before the introduction of the robotic devices, revealed many of the same concerns, with more interest in interactive "social robots" such as Paro and Pepper, particularly from female respondents. Despite a few positive comments about care robots in general, the majority were critical or at least skeptical of lifting robots such as Hug. These survey responses provided several reasons for this "unease" (*fuan*, the opposite of anshin)—a term that care workers used repeatedly to describe their feelings toward Hug. In interpreting these responses, it is useful to refer to the distinction drawn by anthropologist Nicholas Sternsdorff-Cisterna in his analysis of food standards in Japan after the Fukushima nuclear disaster of 2011, between the two similar Japanese words *anzen* (safety) and *anshin* (peace of mind, a psychosocial sense of security). As he notes, "*anzen* speaks to a system based on rationality and consistency in its standards. This is underscored by the fact that *anzen* works as an adjective to describe a condition of being. *Anshin*, however, speaks to questions of the heart" (Sternsdorff-Cisterna 2015, 458). Although care staff were told by Fuji managers that Hug was safe to use, their responses revealed a lack of anshin.

Care workers' concerns, expressed both in the survey and in individual interviews, included the physical movement and comfort of using the robot ("because [robots] can't make fine movements, it's a worry for things like transfer"; "it wasn't comfortable to ride"; "my underarms and chin hurt"), and fears that they might end up depending on robotic devices too much and losing their own care skills ("I think I'm anxious that [we'll] get to depend too much on robots, and the quality of our own care will fall"; "I worry that staff will become dependent on robots"). Many staff members argued that use and acceptance depended on the differential abilities and views of residents, once again highlighting care recipients as individuals rather than the generalized and abstracted "elderly" often referred to by robotics engineers or in government policy documents on the aging population.

Most care workers also expected that using Hug would require a great deal of time and, somewhat paradoxically, effort: "It may be physically more comfortable, but there's no time to use it in a leisurely way"; "Some tasks could probably be reduced, but thinking of the process of using it makes me think it's a waste of time"; "I have the sense that things that will reduce the burden on care staff will take effort, so I feel like in the end we'll stop using them." Another stated, "Transfer robots are big and it takes time to prepare it every time you use it. We can't use it in our current work routine." One care worker even wrote in a survey response that Hug was heavy and itself required transfer—using the same word as that used for the transfer of residents—which introduced a new duty for staff. The idea that Hug would reduce the burden on care staff, referring primarily to moments of physical exertion, yet increase the "effort," referring mainly to time, problematized the idea of robots as straightforward labor-saving devices and suggested that care workers prioritized having more time and "leisure" for care over reducing physical exertion in lifting.

The time taken also related inherently to spatial problems. Matsui told me that at the previous care home where she had worked, they had trialed the use of a sling for lifting but encountered a problem in using it to transfer residents into and out of the communal bath. The process took too long, and residents would get cold while they were being hoisted. Moreover, the layout of the communal bath at Sakura meant that it was simply impossible to use Hug there. Matsui finished her story by noting that such lifting devices could become a hindrance or obstruction if the built environment of the institution into which they are being introduced was not physically designed to accommodate their use.

Finally, possible negative reactions of residents to Hug were brought up as a concern. The idea that using a machine could remove the sense of shame or reserve that residents might feel in asking female staff in particular for help to be transferred, especially if they were aware that it could cause injury, was outweighed by other considerations. Otsuka compared Hug to a forklift. Punning on the words *nimotsu* (luggage) and the more metaphorical *onimotsu* (burden), he told me, "Hug is comfortable and interesting, but I wonder if service users feel like, 'Am I luggage?'" Significantly, as described above, Otsuka was one of the three care staff who had suffered a herniated disc; according to the government's rationale about care worker back pain, he might have been expected to welcome the introduction of Hug. Another member of staff wrote:

> To begin with, using robots, in relation to doing care, if anything my feeling is almost opposed to the idea. I still have the feeling that [care is] after all about people caring for people, and it's disrespectful to our elders.

A different care worker wrote:

> I wonder whether it is appropriate to use robots when a person requires support due to physical problems. In the manufacturing industry it is common to use robots, but in this industry, it has to be people helping people. It's disrespectful toward those who have helped build Japan. Personally, if my grandmother or grandfather were to be taken care of by a robot, I wouldn't like it.

The relation between respect, care, and manual effort was further highlighted by other respondents:

> I think [caregivers] with back pain can use transfer robots, but because I like connection and interchange between people, and am doing this job, I feel like I don't want to rely on robots. I want to assist with my own hands.

As social policy scholar Ishiguro Nobu (2018) notes, the idea of care with human hands is a cultural ideal in Japan, associated with warmth and relationality. Notably at no point during the trial did any staff member express worries about robots taking their job—a widespread concern in North America and Europe. During interviews, care workers explained that they did not feel threatened by robots primarily because of the huge labor shortage in the care industry.

These concerns about Hug, and the general sense of unease among staff, were largely disregarded by the characteristically pragmatic Mr. K, who told me that he thought staff resistance was the result of lack of education, together with the age of care workers (their average age at Sakura was forty-four), and a generalized aversion to change. For him, as for Matsumoto quoted above, successful implementation was a question of getting them used to the devices. In particular, he argued against the claim of some care workers that Hug would not ensure anshin for residents, noting that staff often pushed residents around too quickly in wheelchairs to meet the demands of the institutional schedule:

> I think if I put you in a wheelchair and I wheeled you around, especially at the speed they wheel them around on the third floor, would you feel any anshin? I've seen these people [i.e., residents]—they grip the sides like, "Holy shit, you're going to crash me into a wall any second." Even though they're eighty-eight years old, ninety years old—the way they speed them around like they're F1 racers on those wheelchairs on the third floor—you've seen it![9] You've seen their faces. You think they volunteer for that shit? They don't volunteer for it, that's because that's the world that they live in, and they gotta get used to it. Wheeling them around at lightning speed at Mach 3 on the wheelchairs—I don't think that's anshin! So if you say, "Oh, we can't use the Hug machine because they don't feel anshin," but we can wheel them around at lightning speed just because they don't scream bloody murder, if we're beating on the anshin drum thing. . . . And if we go to a no-lift policy which exists in other countries, then it has nothing to do with whether you feel safer or not—it's the law.

As a result, he pushed ahead with a pilot of Hug, and a trial model was borrowed from Fuji for a period of six weeks from the start of April 2017. Initially, Hug was demonstrated to care staff at daily meetings, and they were encouraged to try it for themselves before it was used on several residents.

At first, the attitude of staff toward Hug, repeated more or less verbatim by several care workers, was, "We won't know till we try it." In actual use, however, the physical realities of individual residents problematized the apparently universal usability of Hug that had been emphasized in its advertising materials and by company managers. A female resident without the use of one hand, who could only use her other hand to hug, started to complain of pain as she began to be lifted, so the procedure was immediately stopped. Another resident also complained of slight discomfort. A third said that she felt comfortable, and after being lifted by Hug a number of times over the next couple of days, she said she had become used to it. Users' differential weight, ability to grip or stand, body shape, and other physical characteristics, as well as their level of fear of being lifted by a machine all seemed to result in different degrees of comfort or discomfort in using Hug.

After several days, Hug had been more or less relegated to the worker room. Care staff said that Hug had not been well received either by workers or residents, with several complaining of discomfort or slight pain under the armpits as they were lifted. Those residents who were hardest for staff to lift, for example those who were partially paralyzed or unable to grip or support themselves at all, also seemed least able to use Hug. As Chiba told me, "Basically, if we can't use it with those people who don't have strength or are paralyzed, then we won't use it." Staff, however, based these conclusions on only a handful of uses of Hug by residents, several of whom said that it was comfortable to use. One notable advocate of Hug was Diego, who used it regularly, particularly during the night shift. Yet, most of the other care staff brought up problems of lack of time and the effort required to use Hug as reasons not to test out the machine further. As one care worker said when asked about using Hug, "I just had no interest in it."

The Rejection of Hug and the Ethics of Care

Was the unwillingness to embrace Hug simply reflective of a general conservatism in Japan toward the adoption of new technologies—akin to the continued preference in certain settings of handwriting and the fax machine over email and messaging apps? Care homes are, of course, already sites of intensive use of biomedical technologies. Almost every aspect of life at Sakura was technologically and medically regulated and mediated to some extent, from food (prepared and served according to the instructions of a professional on-site nutritionist) to medicine to bowel movements (carefully monitored and recorded by care staff and reviewed by nursing staff) to movements across space. The latter took various technological forms: elevators, wheelchairs that could be self-wheeled, wheelchairs that could only be pushed by a care worker, walkers, walking frames and walking sticks, beds that could be electronically adjusted to sit the user up, and even chairs, which care staff often pushed forward under tables as a way to gently restrict the movements of residents with dementia who might fall over if they tried to get up. Lifting involved transferring residents from one technological device (wheelchair, toilet, or bed) to another. In this environment, electronic lifting machines could be seen as just another technology for moving bodies through space.

The intended transition of lifting from bodily technique to robotic technology appeared to be a failure, however. Beyond the mixed results of the initial uses of Hug, the majority of care staff were clearly not keen on the machine to begin with and did not seem interested in trying to make a success of its implementation. Yet, this did not appear to be due to a generalized aversion to technology. Care staff were curious about and interested in the robotic devices and, as we will see in chapter 5, immediately started using Paro with residents on a daily basis without much prompting. Similarly, other high-tech devices, such as iPads for recording medical notes and large wheelchair-accessible bathing machines, had been quickly adopted without significant resistance when they were introduced in previous years.

In her 2018 study of Japanese care robots, Ishiguro Nobu explored negative attitudes among care staff toward transfer devices like Hug. She characterized them as an initial reaction attributable to the fact that the relationship between care workers and the older people they care for is premised on *amae* (broadly translated as a relationship of dependence). She also showed how such attitudes, based on the view that using the devices means treating older people like objects (as Otsuka put it, like "luggage"), could change once these devices started being used regularly—a stage in their introduction that care workers at Sakura never reached with Hug.[10]

Certainly, the seemingly practical rationales of not having time and being too busy to use Hug were more complex and intangible than they may at first appear. Using Hug did indeed take additional time—and space—to wheel around, position, operate, and store; in fact, wider-scale implementation of Hug would seem to create *more* work for care workers rather than saving labor. These rationales, however, implicitly appealed to national discourses about the care labor shortage while providing space for individual staff members to make their own decisions about what kind of care to give, as well as enabling tacit resistance to the imposition by management of new technologies or routines that were perceived unfavorably. Although using Hug would have added time to transferring residents, care workers also frequently took time or made time to talk to and joke with residents in as unhurried a fashion as possible. In fact, on average, when not using Hug, the actual manual lifting element of these interactions with residents took about fifteen seconds out of a total of around five minutes per resident for a typical transfer from bed to wheelchair. The rest of the time was spent joking, helping residents change clothes, checking their mobile toilets, and exchanging the kind of "empty gossip" described by Iwasaki in chapter 3—what staff characterized as *yutori* (leisurely), and necessarily personalized, care. In part, the insistence on labor-intensive manual care may have been a way to protect the "inefficiency"—and quality—of care.

Use of Hug with every resident would have meant restructuring the nagare of daily care practices such as waking residents up, serving them meals, and taking them to the toilet. This would have been a somewhat disruptive yet feasible adjustment, but it would also have reduced *yoyū*—the "surplus capacity" that could be spent with individual residents while still completing the institutionally set tasks. Using the rationale of "no time" seemed a way to conceal other reasons for rejecting the technology as well as preserving the status quo. Through the use of such rationales, care workers at Sakura, despite holding jobs that are perceived as relatively low status in contemporary Japanese society, perhaps surprisingly also held a significant degree of power over the adoption of robotic devices at the home, particularly given Mr. K's inverted-pyramid style of management and the high degree of labor mobility in the industry. This suggests one reason why the government's strategy of promoting robotic devices directly to managers in the care sector and providing financial subsidies for care homes to purchase such devices was unlikely to be sufficiently effective to guarantee their widespread use.

Moreover, it was clear from speaking to staff and from observing transfers taking place that lifting constituted an integral part of the intimate care delivered by care workers. Transfer provided close bodily contact as well as being part of a routine of talking, joking, and performing everyday sociality in order to "get close to" residents. In fact, caring with one's own hands, human to human, was identified by staff as perhaps the single most important characteristic of good care, expressed as respecting the elders. Touch played a central role in establishing and communicating peace of mind (anshin), working together with the use of kinship terms, informal language, and joking to cocreate a close familial relationship between caregivers and residents expressed in the idiom of filial care. Yet, it also acknowledged the co-vulnerability inherent in the bodily nature of care practices—a vulnerability that cut both ways. Although Hug may have been, for some staff and residents at least, anzen (physically safe) without providing anshin (peace of mind), the reverse was true for manual lifting.

Anthropologist Jason Danely argues with regard to Japanese informal carers that "by learning compassion, [they] are able to construct meaningful narrative subjectivities that transform personal suffering into the basis for connecting to others and to transcendent or transpersonal modes of encountering the world" (2016, 178). Among Sakura's professional care workers, compassion was similarly important, and the "co-suffering" (Danely's term) embodied in the performance of transfer, including the back pain that might result, was accepted by many (though not all) care staff as part and parcel of the job. Manual transfer as a bodily technique for communicating and doing care provided symbolic meaning to the physical act of transferring a patient and established the value of care labor. If residents were reduced to mere objects to be moved around like "luggage"—something closer to the engineering-centric views of some researchers at RIRC who tended to conceptualize older people as abstracted, relationless, parameterized objects—then care workers would become manual laborers in a warehouse. Lifting added weight to an ethically embodied care, while acknowledging the material substance—the corporeality—of residents.

Within the context of an underlying ethic and ideology of personalized care, technological mediation of this social and tactile care relationship was perceived by staff at Sakura as disrupting the "connection and interchange between people" that most of them said they valued. The nature of the "disrespect" from using robots to care involved not being able to respond to older care recipients as individuals both physically and socially. Physically, in terms of respecting individual bodily peculiarities that did not answer to generalized and abstracted textbook models of aged bodies and adjusting tactile contact according to their perception of older residents' affective reaction to touch. Socially, in terms of treating each resident respectfully with yoyū (surplus capacity of time/space) as an individual with their own personality. In threatening to replace yoyū with additional institutional tasks, Hug threatened the ethical basis of care. In a sense, then, its use seemed not only to disrespect older residents but indeed for some care workers, in an almost symmetrical manner, to diminish the value and dignity of their own work and the pride they took in it.

5

PARO

Reconfiguring Communication

Paro is a robot shaped like a seal pup. Weighing in at six pounds and measuring around twenty-two inches in length, it resembles a large plush toy (see figure 3). Marketed as a therapeutic communication device, it has been developed by Dr. Shibata Takanori at the National Institute of Advanced Industrial Science and Technology (AIST) and Massachusetts Institute of Technology (MIT) since 1993, tested for use with older adults since 2000, and was eventually released for sale in 2005. This makes it one of the first robots to have been used for care—comparatively ancient compared with more recent models like Pepper—although it has undergone several updates since it was first released. Over this time, it has become probably the best-known care robot, both in Japan and worldwide. Overcoming some of the critiques of other service robot designs being cold and hard, Paro is warm and tactile: it has soft antibacterial fur and an internal regulator that keeps it at human body temperature and responds to sound and touch by moving its head, blinking, and crying like a seal. It is also supposed to react to the sound of its name and a few words including simple greetings and compliments. The form of a seal, rather than that of a cat or dog, was chosen because the user was expected to be unfamiliar with a seal and harbor fewer expectations of its appearance, sound, or behavior—it is intentionally designed with novelty in mind. Paro is also branded somewhat ambiguously as "mental commitment robot Paro," suggesting a certain burden of responsibility on the user—either to commit to therapy or, perhaps, to the formation of a relationship with the robot.

Every Paro unit is superficially unique, crafted through a process of hand finishing in Shibata's hometown of Nanto City. As Selma Šabanović notes, each unit

FIGURE 3. Paro. Photo by author.

is given a birth certificate, and the manufacturing process exemplifies "attention to detail and quality.... Paro's design embodies cultural models of skill, quality, relational construction of value, and appreciation for local tradition in emerging robotic technology" (2014, 349–50).[1] Paro is, in this sense, a personalized robot, reflecting its name—a portmanteau of *pāsonaru robotto*, or "personal robot." Yuji Sone argues that Japanese users value this veneer of personalization: for example, "AIBO [Sony's dog-shaped robot] owners are attached to their robots and often sustain a belief in the singular uniqueness of their machines" (2017, 207).

Šabanović describes how Shibata attempted to link Paro to traditions of Japanese craftsmanship and an emerging Japanese robotic culture, embodying certain supposedly archetypical Japanese values. Yet as in the case of Pepper, Paro's origins are not straightforwardly Japanese. In an interview I conducted with Shibata in 2016, he explained how, back in the 1990s, managers at AIST were not able to see the potential of his invention. This was a time when publicly funded robotics was coming to be seen by some as a collection of "useless" pet projects and a waste of taxpayers' money: "AIST didn't understand the purpose of my research, so they couldn't... they didn't give me any budget related to Paro [laughs]. That's why I moved to MIT from 1995 to 1998." At the time, engineers at MIT were pioneering facially expressive and socially interactive robots such as Cynthia Breazeal's Kismet robot head that were more attuned to Shibata's research. In terms of dissemination, too, Paro has been relatively more popular outside of Japan than domestically—it has been exhibited internationally and has sold fairly well in Northern European countries—and Shibata has made ongoing attempts to promote Paro abroad, a general trend across much of Japan's care robotics industry, spurred by bureaucratic obstacles and slow sales in Japan itself.

By 2018, Paro had cost approximately $20 million to develop, and a rather modest total of around five thousand units had been sold across thirty countries, with many of these sales in Northern Europe. At the time of writing, it was being marketed for sale in Japan for ¥453,600 ($4,500) with a three-year warranty. Shibata stated that his inability so far to get Paro confirmed, in Japan, as either a medical device or a device to be included in the Long-Term Care Insurance system, where it would receive a 70–90 percent subsidy from the government, may have contributed to the relatively low volume of sales.[2]

Shibata told me that he had developed Paro to serve two purposes:

> One was [to be a] personal robot, the other was [to provide] animal therapy. I thought we don't need a human-type robot at home. But we have pets. So that means, I thought I should develop some machine that is not for a task—so I should develop something useless in terms of work. I wondered what we have in our lives, and pet animals are something like that. So we don't expect pets to work for us, but still we want them, and we love them. And I wondered what kind of role they have. So pet animals can't do some tasks, but they can enrich our lives psychologically.

Paro was thus not designed exclusively for eldercare, nor was this initially the primary use for many of the units sold. Shibata has stated that "most Paros in Japan were not sold for care purposes. In the Tokyo area, many condos and apartment buildings do not allow pets, so many people bought the robotic seal as a pet substitute" (quoted in Kohlbacher and Rabe 2015, 36).

Shibata has also long argued that Paro could act in an analogous way to animals in therapeutic roles without the drawbacks of a real animal's germs, hair, unpredictability, maintenance needs, and mortality. In 2002, even before it was available for sale, Paro was awarded the title of "the world's most therapeutic robot" by Guinness World Records.[3] When I asked Shibata what was meant by the term *therapeutic* and how this was translated into Japanese,[4] he replied,

> In Japan, the word or term *therapeutic* has some vague kind of meaning. So depending on the purpose of explaining Paro, I choose some different kind of words. So for the public, *iyashi* [healing] has kind of a broad meaning, but it doesn't have a clear definition. As for *therapy*, it's for like a medical service—so at that time we needed to show some kind of evidence. So yeah, depending on the purpose of the application or explanation of Paro, I needed to choose the right word.

He acknowledged the subjective nature of evaluating robots against these different definitions, later remarking, "As for the robot itself, as humans interact with robots, the human evaluates the robot, or interprets the meaning of the robot subjectively. So it's very difficult to evaluate the robot by objective measures."

The Guinness label of "the world's most therapeutic robot" may have seemed a gimmick since there were very few competing candidates for the title at the time and little evidence to support such a vague claim. Nevertheless, the superlative power and novelty value of the award seemed to have a beneficial effect on Paro's international reputation. Paro is certainly the most studied care robot. Over the past two decades of its circulation in care homes, it has constituted a frequent object of interdisciplinary academic studies and media attention that far outweighs its actual commercial popularity or usage. An International Symposium on Robot Therapy with Seal Robot, Paro has been held nine times in Japan since 2012, sponsored by organizations such as MIT's AgeLab and Oxford University's Institute of Population Ageing. Various claims have been made regarding Paro's benefits for older users, particularly those with dementia: that it can alleviate stress, improve mood, reduce feelings of loneliness and depression, increase social interaction with others, reduce the need for psychotropic medication, and reduce pacing or wandering (Santos, Yoon, and Park 2015).[5]

As well as sparking a huge amount of scholarly interest in social robots and providing a key reference point for the incipient field of human-robot interaction studies, Paro, together with AIBO (Sony's dog-shaped robot first released in 1999), contributed to a boom in the manufacture of interactive robots in Japan and beyond, particularly those shaped as animals, small children, or soft toys. Through the work of Shibata and others, these began to be seen not simply as toys but as therapeutic devices with potential benefits for health and well-being. During my research in 2016, for example, all three of the main telecoms carriers in Japan had launched their own communication robots: SoftBank's Pepper, KDDI au's Comi Kuma, and NTT DoCoMo's Kokokuma. The latter two were both shaped as soft teddy bears; other toy-like robots included PIP&WiZ's Kabochan, Fusion Marketing's Orikō Kuma-tan, a plastic teddy bear–shaped device that could play songs and tell stories, and Groove X's Lovot, which was created by the former lead developer of Pepper and was the recipient of an unprecedented $70 million in early-stage venture capital investment.

Yet inside AIST itself, there seemed less regard for robots like Paro. As Hirukawa, the head of intelligent systems, tweeted:

> When it comes to robots, it's a mystery why many people are thinking about communication robots. Would you really pay ¥500,000 [$5,000] for communication? (@hirohisah, February 23, 2016[6])

As my fieldwork at AIST has indicated, communication seemed less valued among engineers involved in robot care. Indeed, the fairly frequent description in press releases and in the media (and even in Shibata's own words quoted above) of social robots like Paro, AIBO, or Lovot as "useless" suggests the devaluation and discounting of communication, social interaction, or any other attributes that do not appear to have an immediate productive function.[7] This echoes the common assumption that human communication skills in relation to care work are gendered and innate and require neither training nor financial compensation. But as one engineer at the Robot Innovation Research Center told me, "One of the big problems for elderly people—they have too much spare time. They need to fill such spare time with something. Social robots might be used in such cases." Eldercare appeared to provide "useless" robots with a commercial purpose.

Questions have been raised, particularly by Euro-American scholars, about the ethics of using social robots like Paro with older users. Their design and use have been characterized as deceptive and manipulative, particularly in relation to those living with dementia, who may find it difficult to separate the fantasy presented by the robot (that the user is interacting with a sentient being that might empathize with or care about them) from the "reality"—that they are simply interacting with a machine that feels and understands nothing (Turkle 2011; A. Sharkey and N. Sharkey 2012). Social robots have also been represented as depersonalizing: degrading both the quality of social interactions and quantity of human-human communication, as interacting with the robot is assumed to replace at least some degree of interaction with other people (Sparrow and Sparrow 2006). Sherry Turkle (2011) argues that social robots can leave us feeling "alone together," or increasingly isolated through ersatz social interaction with, or mediated by, machines. Similarly, Kathleen Richardson sees a growing "attachment crisis" in how people relate to each other, which she claims results in the "mechanistic sociality that underscores contemporary sociality. The mechanistic sociality is an outcome of an attachment crisis in how humans bond with others . . . in accepting the mechanical robot, humanity must become more socially mechanical—less complex, more scripted, more stereotyped, and less spontaneous, adapting to the needs of the machine" (2015, 131–32). Judith Donath (2020) likewise argues that interacting with artificial entities such as robots reinforces a way of communicating that is increasingly instrumental and one-way rather than nurturing and empathic.

At the same time, long-standing anthropological and science and technology studies interest in human relations with animals and other nonhuman entities, as well as more recent work on posthumanism, challenges the idea that human-human interactions are exclusively authentic, nondeceptive, and "natural." Some scholars have suggested that "social" robots such as Paro exist for users at times as nonhuman beings that are cathected with emotions, social meanings, and values.

In other words, they argue that Paro can have as much of an inner life and authentic connection to us as we are willing to give it. Moreover, some argue that rather than thinking in terms of deception when it comes to social robots, we might instead think of users of such robots as suspending disbelief in a similar way to enjoying a movie or playing a video game (Donath 2020).

Others have deconstructed the assumed binary of cold technology opposed to warm human care, arguing that technology is not innately "cold" nor human care necessarily "warm" and that in any case, the two can never be neatly separated out (Mol, Moser, and Pols 2010). Darian Meacham and Matthew Studley go further, arguing from a phenomenological perspective that "what matters in a caring relation is not the internal states of the agents participating in the relation, but rather a meaningful context: a care environment that is formed by gestures, movements, and articulations that express attentiveness and responsiveness to vulnerabilities within the relevant context" (2017, 98). They argue that if robots could be used to construct such a caring environment, they could care just as well as—or better than—human caregivers. Philosopher Paul Dumouchel reverses Turkle's critique, arguing that being "alone together" with a robot can in fact prove beneficial in a communal institutional environment where residents have very little privacy. He argues that Paro can provide an intimate one-to-one relationship that raises the quality of life among users by transcending the "socially impoverished" institutional setting (Dumouchel 2016; Dumouchel and Damiano 2017). Some have also suggested that robots such as Paro can be used as a social tool in the care home, to act as a new medium of communication between residents and staff, and therefore improve the quality of human caring relationships.

Mr. K's hopes in introducing Paro were somewhat different. He said that a major problem at Sakura was short-stay residents who suffered from severe dementia and therefore tended to take up a lot of staff time through repetitious requests. He said that Paro would be ideal to use with these residents to keep them occupied, settle them down, and give staff more time. Regulation of residents' affectively expressed moods was an important goal. During my fieldwork, I often saw care workers trying to mollify angry or loud residents; this often meant spatially separating them from other residents and staff and giving them time to calm down. It could also mean administering medicines such as Aricept (donepezil), a drug used for controlling the behavior of residents with dementia, who might sometimes scream, bang on the tables, or lash out. It was hoped that using Paro might reduce the need for such medications. Mr. K told me that he had considered animal therapy, but although there was no law prohibiting bringing a dog into the care home for residents to pet, if just one resident was scared of dogs or had an allergy, they would not be able to use it because "how would you control the [care home] floor?" He was therefore excited about the possibility of Paro providing the same benefits without the hassle of a real animal and while maintaining "control" of the institutional environment.

Societal Crisis and Therapeutic Communication in Japan

The concepts underlying Paro have deep roots in a broader intellectual ferment of ideas about therapy, communication, and their digitalization—and, in particular, about *iyashi* and *keichō*. In the field of Japanese studies, the term *iyashi* (healing or relaxation) has been used as a key frame of reference for understanding social robots like Paro. Discourses of iyashi arose in the 1980s, imported from the U.S. New Age movement. They were promoted more aggressively by Japanese media and corporations in the wake of the Kobe earthquake and attack by the Aum Shinrikyō cult on the Tokyo subway in 1995 as a way to cope with the economic stresses, problematic human relations, and natural and manmade disasters of post-bubble life.[8] Iyashi is one of a number of therapeutic discourses loosely connected to the idea of *kokoro no kea* (care of the heart). Another important related practice, which also emerged following the events of 1995 and again after the 2011 Japanese earthquake and tsunami, was keichō (usually translated as "active listening" or sometimes as "listening volunteerism"). Following these disasters, there was a sense of large numbers of victims suffering in silence, with no outlet to express their grief, anger, or stress; the aim of keichō was to listen to these experiences and through listening, to help heal them. The spread of therapeutic techniques and technologies of mental health also coincided with a growing public recognition of and discourse around the rising numbers of older people in Japan suffering from dementia.

On a national scale, the mental health "crises" precipitated by natural or manmade disaster and older age seemed symptomatic of broader interconnected and ongoing challenges of post-bubble Japan itself: the lost decades of economic stagnation, the aging population, and unfolding forms of neoliberal political economy that were atomizing society and promoting economic and social precarity—in other words, as with the related crisis of eldercare, crises of Japanese capitalism itself. These have often taken the specific form of a breakdown of communication and social relationality, linked to various problems such as (mostly male) socially withdrawn youth who spend most of their time in their rooms (*hikikomori*); increasing, often hidden, homelessness; and lonely older people living and dying by themselves. Communication breakdown appeared to substantiate the emerging image of twenty-first century Japan as a *muen shakai*: a society where the bonds of sociality have come untied. Whereas the communication provided by "useless" robots may have been discounted by some engineers at AIST, in broader Japanese society, communication was increasingly being seen as synonymous with practices of care in areas of mental health and dementia—a way to heal society.

Yet, these healing practices are also disputed. Media studies scholar Paul Roquet (2009) suggests that iyashi (healing) has been employed opportunistically, at a time of socioeconomic restructuring during which precarious employment has soared, as a kind of individualistic self-help technique that by treating the damaging effects of neoliberal political agendas also helps sustain them. Although keichō (active listening) may be an ethically informed, humanitarian practice motivated neither by profit nor productivity, some also see it as a direct expression of neoliberal governmentality, helping delegate responsibility for mental health and welfare to individual volunteers or family members.

"Healing," Paro, and Affective Time-Space Escapism

Roquet argues that iyashi was just the latest in a series of "healing booms" and part of a long-term "shift towards technologies of mood regulation" in Japan since the late 1970s (2009, 88). The term *iyashi* has been applied to a wide variety of goods, services, environments, and people, including music, travel, toys, robots, cat cafés, adult film actors, Nobel Prize winners, therapies, novels, and more. As Roquet notes, the common thread is that all are said to have a healing effect, often invoked through their sensual affective properties, creating a "transposable" or "portable affective calm" that establishes peace of mind (anshin). Affect here refers to a precognitive sensory response—a sensation without thinking—in contrast to emotion, which involves interpreting how you feel, labeling it, or putting it into a category. Iyashi fosters a calm and relaxed affective state as an end in itself.

By the time of my fieldwork, iyashi was a characteristic routinely ascribed to social robots. For example, during an interview with PIP&WiZ, a company that makes Kabochan, a small soft-toy-like robot shaped like a little boy, a manager told me that "just by being there for you, Kabochan heals." In the context of eldercare, this "healing" applied to both caregivers and recipients of care: the company representatives told me that it could relax service users and help heal their presumed loneliness, while equipping caregivers with "surplus capacity in their hearts" (*kokoro ni yoyū*)—the same phrase used by care workers at Sakura to describe a key requirement for providing good personalized care. One of the ways in which robot engineers claimed communication robots could "heal" was through their contextless, simple affect, and sensual tactility. In stark contrast to the complexity and sociocultural parsing of communication through filters of gender, seniority, formality, and situation in everyday life in Japan, communication robots are often modeled as animals or young children—to mimic a naivete and cute (*kawaii*) ingenuousness lacking in adult relationships. When you cuddle a warm, fluffy, purring robot, you do not need to think—you can simply relax and let yourself be "healed" by the sensations of the interaction.

The concept of iyashi is bound up with notions of subjectively experienced space and time: the aim of iyashi products is to evoke an affective response in order to transport the user out of their everyday routine. Roquet notes that one writer of iyashi novels, Kurita Yuki, describes how her work provides space (*ma*) for the reader, and he relates this to Gilles Deleuze's description of "amniotic spaces": "we no longer know what is imaginary or real, physical or mental, in the situation, not because they are confused, but because we do not have to know and there is no longer even a place from which to ask" (Deleuze 1989, 7 cited in Roquet 2009, 103). In her analysis of cat cafés in Tokyo, Lorraine Plourde describes how iyashi interactions with cats involve the staging of an affective domestic space, where the postwar ideology of the productive family unit is rejected and the customer can relax and "heal" in a nonwork environment "positioned outside the frenzied and presumably stressful experience of everyday life in Tokyo" (2014, 122). The creation of such nonproductive time-spaces—suffused with a sense of stability and security—involves a relaxation of boundaries between real and imaginary, self and nonself, as one loses oneself in the moment. This provides an alternative way of understanding the "deceptive" nature of social robots like Paro as integral to their functioning as iyashi.

Affectively escaping from productivist Japan has its price. Both Roquet and Plourde highlight the commercialization of affect in the iyashi industry: Plourde sees the phenomenon of iyashi as part of a growing immaterialization of the postindustrial Japanese economy in which social relations are increasingly transformed into products and services available to those who can afford them. Nevertheless, both also argue for its redemptive therapeutic potential. But withdrawal from the sociopolitics of the world and human relationality into the "amniotic" womb-like self—in other words, escaping from the reality of one's everyday life and complex social relationships—can also create other problems. Iyashi is concerned with withdrawal from the world into a subjective, affective spatiotemporal bubble but risks becoming a permanent and detrimental escape from reality—a self-absorbed turn to inner life as a way to avoid confronting trauma or uncomfortable social realities.

"Active Listening": Communication as Therapeutic Mirror

As iyashi gained popularity, the related practice of keichō (active listening) was also beginning to enter the mainstream. The term *active listening* was originally coined in 1957 by American psychologists Carl Rogers and Richard Farson to describe a communication technique that could be used in daily life and business. At a basic level, active listening involves listening with care to your interlocutor in a one-on-one environment and mirroring back to them what they say. It was based on a form of psychotherapy pioneered by Rogers since the 1940s in which the therapist acts as "a caring and congruent mirror," reformulating and repeating back what their patient tells them (Prochaska and Norcross 2002, 133–62).

The Japanese therapeutic practice of keichō is directly based on active listening. The listener reformulates and summarizes what their partner says without introducing any personal opinions, judgments, advice, or expressions of emotion beyond smiling in an encouraging way. In order to learn more about keichō, I completed a one-day training course in February 2017 run by a nonprofit organization in Tokyo.[9] The instructor explained, "First, you have to listen to what the person you are talking to is saying and accept their worldview.... You have to set aside your own feelings and react to and repeat what the other person is feeling right now." The person speaking is encouraged to talk about whatever is on their mind, and the listener can also ask open questions to elicit further responses. The practice is based on the idea that everyone has a desire for positive regard or validation, which is served by being heard through the active performance of listening. At the same time, keichō practitioners claim that using the technique in everyday situations also transforms their own lives. Just as the 1957 article argued that the practice of active listening has the potential to make one more effective in business, so too does keichō offer the opportunity to become more successful in every aspect of social life.

Keichō was carried out regularly at Sakura by the Teddy Ears volunteer group, consisting of fifty women and two men in their sixties and seventies based at the local community health center for those over sixty years of age.[10] The name of the group is a pseudonym but seeks to capture the somewhat infantilizing nuance of the actual name, with the tactile image of a soft animal's large ears—an animal that, importantly, is unable to talk back but can act as a listening companion: precisely the role that Paro was also expected to play. Usually, one or two volunteers from the group would visit Sakura twice a week and spend around half an hour on each floor. Doing keichō "properly" in the setting of Sakura was extremely challenging since it relies on a partner who wants to, and is able to, speak. Despite the theory of active listening, it was usually necessary for volunteers to do most of the talking. It was difficult to listen both because residents were often quiet and did not talk and because thirty minutes was a short time to speak with two or three residents. Sometimes this involved general conversation about where the resident was from, their background and interests, and sometimes it involved the volunteer asking simple math questions such as multiplication tables. As a result, it seemed to represent less an "active" type of listening than a proactive form of communication.

When I asked the group about this and about the broader difficulty of doing keichō with people who suffered from dementia, one member agreed that there was a gap between how they were trained to do keichō and how they actually did it. She said, however, the most important thing was that older people "respond to things like touch, eye contact, and a smile." Other volunteers emphasized the need to treat and respect every person as an individual. Another added that initially, people's hearts were closed and that keichō was about opening up and making a connection between their hearts. In this way, keichō in actual practice constituted a degree of listening and giving positive regard to residents while also serving as a starting point for initiating and building a social connection between volunteers and residents—a practice of bringing yoyū, or the "capacity" for slow communication, into the institutional routine, mirroring the ethics of personalized care shared by care workers.

Yet, keichō in its ideal form is a robotic method of communication. From the perspective of the speaker, the listener is radically depersonalized, acting purely as a positive and accepting conversational mirror to the speaker. Like iyashi, it provides space outside the everyday, enabling the speaker to assess what he or she is saying and thinking from the critical distance gained by projecting their self onto someone else. Indeed, the instructor of the keichō course I attended told me that "keichō is iyashi." The listener does not critically analyze or judge what the speaker is saying but simply accepts it. In this way, the practice operates similarly to several technological forms of therapy that were, in fact, directly developed out of the same intellectual tradition.

Computerized Communication Therapies

In their landmark analysis of which jobs currently done by humans were most likely to be automated out of existence, Carl Benedikt Frey and Michael Osborne (2013) strikingly ranked recreational therapists at number one, the safest, out of a list of 702 occupations, while several other types of therapists and psychologists are ranked within the top twenty-five. Therapy tends to be regarded as human labor intensive and not easily amenable to automation. Nevertheless, computerized and roboticized forms of therapy have long been a prominent area of research and have gained greater momentum in recent years.

The first attempt to automate therapy was a joke that accidentally established a research field. At MIT in the 1960s, computer scientist Joseph Weizenbaum was developing the first chatbot, a program called ELIZA. He created a version of the software, DOCTOR, to "parody" the role of a Rogerian psychotherapist using active listening. Students on campus could type in their feelings and problems, and DOCTOR would rephrase it and repeat it back to them, often as a question. Expecting to demonstrate the superficiality of computer-generated communication, Weizenbaum was instead shocked to discover that DOCTOR's users, despite being told that the computer program did not understand what they were saying, spent hours typing the most intimate details of their lives into the system. These users included his secretary, and as Weizenbaum recounted (in Curtis 2016), "After two or three interchanges with the machine, she turned to me and she said, 'Would you mind leaving the room please?'" Weizenbaum concluded, sardonically, that human proclivity to anthropomorphize computers meant that "extremely short exposures to a relatively simple computer program could induce powerful delusional thinking in quite normal people" (1976, 7). Over time, he became increasingly concerned and outspoken about the dangers posed by computers toward human social relations (Donath 2020, 58).

Sherry Turkle, herself one of the original students who had tested out ELIZA at MIT, characterized these users' reactions as an "ELIZA effect"—"human complicity in a digital fantasy"—arguing that "our willingness to engage with the inanimate does not depend on being deceived but on wanting to fill in the blanks" (2011, 24). Even though users knew the machine did not understand them and were not "really" deceived, they still engaged in a kind of as-if play with it—a voluntary suspension of disbelief. Turkle further argued that increasingly mechanistic understandings of how the brain works have led to greater acceptance of such computer psychotherapy programs.

During my fieldwork at AIST, the reason for the visit of an American researcher was to combine his team's software research on an animated therapist (SimSensei), used with U.S. veterans suffering from post-traumatic stress disorder (PTSD),[11] with the hardware of the lifelike Actroid F androids—in other words, to create an embodied nonhuman therapist in the tradition of DOCTOR. As a keichō practitioner might do, SimSensei asked generic open-ended questions, repeated back responses, and offered encouraging noises, expressions of sympathy, or continuation questions. The team in the United States had found that SimSensei appeared to be more effective than a human therapist in eliciting willingness to open up among their sample of American veterans. The purpose of this researcher's visit was no coincidence, given the increasing prominence given to technological interventions in mental health treatments and therapies—untapped pockets of labor-intensive profit potential, to say nothing of the large numbers of U.S. veterans of the wars in Iraq and Afghanistan who were expected to benefit from therapy.[12]

There are several important parallels between iyashi and keichō. Both practices were adapted from therapeutic techniques developed in the United States; both are concerned with "care of the heart" (kokoro no kea) and healing the self; and both aim to achieve this by transporting the subject into an alternate, decontextualized, subjective state of mind or temporality focused on the self. Although deliberately nonpolitical, both discourses of societal healing flourished in Japan in parallel with socioeconomic and humanitarian trauma but at the same time have come to reify and indeed coconstruct such trauma and "societal problems" of neoliberal capitalism. And while coconstructing the problem, they also coconstruct the remedy: anshin (peace of mind), a state of psychic stability achieved through healing practices of simplified, sometimes mechanistic, communication therapy.

Paro was developed at AIST and MIT within this intellectual ferment and technical tradition. Like DOCTOR, SimSensei, or a keichō volunteer, Paro does not need to understand the user or their problems in order to do therapy because the therapeutic effect comes from the self of the subject—Paro merely stimulates it by engaging the affective imagination and producing an affective response; indeed, the fact that it is not human and cannot understand or judge is an asset.[13] Like other iyashi products, Paro represents commercialized affective interaction, but as a robot, it also reconfigures communication as a mechanically and digitally reproducible, scalable, commodified, and one-sided form of therapy.

Paro in Practice

When Paro was introduced at Sakura in late March 2016, it quickly proved popular, with the majority of residents responding with pleasure to this new creature and reaching out to pet it when it was placed in front of them. The initial reaction to Paro can be summed up by a note that was passed to me by a care worker recording reactions from residents on the first day of its use:

> Care robot Paro; 30th March:
> "Cute" x7
> "Scary" x2
> "It would be better if it could talk"

> There were many favorable opinions. Users gathered around the place where it was being used. Users who don't often show smiles were smiling and they demonstrated the behavior of touching it by themselves.

Care staff told me that residents were "very happy" with Paro and that it had a positive effect on most of them. Several care workers had told me that residents were prone to feeling lonely and indeed that they "felt lonely *because* they are in a crowd," and it was hoped that Paro could ameliorate these feelings by providing a pet-like companion.

Several care workers described Paro as cute, and some seemed enchanted by it. Fujita in particular became a strong advocate of Paro, taking it around all the residents on her floor when she had time, praising it to them, and encouraging them to play with it. She admitted that she was "mad about" Paro, talking to it frequently, mock apologizing to it for leaving when she got up from the table where it was perched, and jokingly complaining that she was "lonely" without it when it was taken to a different floor. At her instigation, staff first gave it the nickname Maron-chan (Chestnut), which was later adjusted to Meron-chan (Melon), in each case using the diminutive suffix—*chan* reserved for children or close friends. Later, care staff and residents forgot these names and went back to calling it Paro-chan.

Kimura was the first resident to be given Paro regularly. She was a favorite of the staff, who characterized her as a cute and gentle grandma. Kimura would sit quietly, periodically rubbing her eyes or face, and slowly raising her face to respond quietly to care workers, who would often sit with her and give her a back rub or ask her to fold towels. In a typical interaction, care staff put Paro on the table in front of Kimura and her neighbor and said, "Here's Paro-chan." Kimura leaned toward Paro and gently called, "Paro-san." Paro mewed in response and wiggled its tail, and soon Kimura went back to rubbing her eyes. Kimura rarely seemed to react to or show much interest in her environment, so it seemed significant that she was focusing on Paro for some time. Over the course of a couple of hours, she spent a few moments stroking or nuzzling Paro to her face or smiling at it and then appeared to lose interest and returned to sitting quietly or napping. Then a few minutes later, she would go back to stroking it and calling to it, "Paro-kun[14] . . . Paro-chan," although Maeda told me that she later forgot the name: "She has dementia, so everything is always new to her!" Occasionally, Kimura would also make comments about Paro to her neighbor, apparently empathizing with Paro as it closed its eyes: "Are you sleepy?" Other residents and staff around her would often interact with Paro, sometimes walking over and asking permission from her before petting it.

Studies of social robots such as Paro have sometimes focused on how the framing and staging of the therapeutic encounter contribute to a sense of novelty that can draw older adult users out of the everyday. Sone argues that there is a "theatrical structure" to robot therapy—it is staged as a performance, "with the facilitator as MC, robots as performers, and patients as the audience" (2017, 197). In Christina Leeson's description of the use of the Japanese robot Telenoid in Danish care facilities, this sense of occasion was evoked through ritualized use of space and time, with special areas of the institutions designated for short periods of Telenoid use. But this practice was not sustainable: the trial of Telenoid in one care home was dropped because it was too labor intensive, even for a single one-hour session per week. Similarly, in actual use at Sakura, care staff did not follow the formal guidelines in the printed user guide that accompanied Paro, which recommended holding specific, time-limited therapy sessions. Care workers said they had no time to conduct hourlong group sessions with Paro—the whole point for them was to give residents something to occupy them so that they could go about their other duties with fewer interruptions. On the other hand, nor did they try to coerce residents into playing with Paro. If residents did not express interest in Paro, they could simply leave it; after some time, it would "go to sleep" and care staff would take it away. They tended to gauge which residents seemed most interested in playing with Paro and then focused on regularly giving it to that resident, who would sometimes also request it. Whereas the use of Pepper and Hug was more closely scheduled, Paro was initially used on a far more ad hoc basis reflecting a "free play" model of use that Mr. K had recommended to his staff.

Paro was used almost daily, although this depended on the care workers on duty. When it was brought out, it tended to be used for a few hours in the morning, usually with two or three residents. At the start of the trial, staff often told me that the great benefit of Paro was that they did not have to keep an eye on it—they could simply put it on a table for residents to play with, and it could keep them somewhat entertained without the need for close supervision or additional labor. As Mr. K told me, in contrast to Hug or Pepper, Paro was "a passive robot . . . noninvolvement—you pass it out and you don't have to do anything." Care workers would often mimic the mewing sounds made by Paro, and I frequently heard them praising it to residents, saying things like, "Wow, isn't it amazing!" Several used the term *iyashi* to describe Paro. For example, Nomura said that because it was cute and pleasant to touch, "in that way, I think it's definitely becoming healing [iyashi]."

But contrary to Mr. K's hope that the use of Paro would reduce the repetitious demands of some residents with dementia, this did not prove to be the case. On several occasions, staff set Paro down in front of the most demanding short-stay resident, Inoue, who would make requests of staff very frequently throughout the day. Yet, Inoue showed little interest in Paro, and it seemed to be ineffective in reducing her repetitive demands of staff or reducing their workload. In fact, using

Paro ended up creating new challenges for care staff. Two residents took a particular interest in Paro. One was a male resident, Takahashi, the former jazz guitarist mentioned in chapter 3 who always wore a beanie hat. Fujita described him as "a grandpa who likes stuffed toys," and indeed he often held a big teddy bear in front of him on the table. His face would light up when he saw Paro, and he seemed to enjoy playing with it. After some time, though, he figured out how to "skin" Paro by unzipping and partially removing its fur, exposing the robotic body underneath. As he started to do this routinely, care staff stopped giving him Paro to play with. Another long-term resident, too, became attached to Paro to the point where care workers started to become concerned about her relationship with the device.

"The Rhythm Goes Crazy": Wandering in the World of Dementia

Around a week into the trial period, I started to hear from several care workers that one resident, Ito, loved Paro and took it everywhere with her. Ito was eighty-five years old, used a wheelchair, and had severe dementia. Iwasaki characterized her as "the type of kind person who would collect stray cats" and added that she had a great love for animals.

Her interactions with Paro followed an almost identical pattern. She would start by playing with Paro, usually with other women seated at her table in the dining room. They patted and stroked the device and talked about how cute it was. Ito would then lift Paro off the table and put it into her lap, before slowly and discreetly wheeling herself back to her bedroom (see figure 4). There, she would put Paro to bed, close the curtain that separated it from the rest of the shared bedroom, and start to talk to Paro quietly, often crying as she did so.[15] At other times, when care workers brought her back to the dining hall, Ito would sit with Paro on her lap swaddled in a blanket like a baby.

Several care workers told me the same story, which I soon observed for myself: Ito stopped talking to other residents, grew uneasy when separated from Paro, and even stopped eating because she worried about Paro, which was taken away during mealtimes. They said that it was dangerous that separation from Paro was affecting her regular communal mealtimes and that she was spatially, socially, and apparently, psychologically withdrawing from other residents. On the other hand, although she spent hours with Paro, care staff were noticeably averse to taking it away from her or giving it to other residents instead.

Similar observations of ethically problematic interactions between older adult users and Paro have been made in other cultural contexts. Phie Ambo's 2007 documentary film *Mechanical Love* presents the case of an older woman, Frau

FIGURE 4. Ito prepares to take Paro back to her room. Photo by author.

Körner, who appears to socially withdraw from other residents of a nursing home in Germany because of what they call her "not normal" affection for Paro. In one memorable scene, she brings it along to her group singing practice, cuddling it in her lap and fussing over it instead of taking part in the session, which Paro continuously disrupts with its loud calls. Another older woman in the group mocks her: "The old woman still plays with dolls." In fact, Frau Körner is unable to sing anyway because she has been talking to Paro so much that she has lost her voice.

The reactions of Sakura staff to these events were ambivalent, as they made clear both in interviews and later questionnaires that I conducted at the end of the trial. Some responded either by pitying Ito or by expressing distaste or unease. Care workers described her behavior as "strange" and "disquieted" and worried that it was making Ito's dementia worse by encouraging her to retreat physically into her own room and mentally into her own world. For example, Matsuo explained,

> Paro has its bad parts. Certainly it can superficially delight, but depending on the person, some think it's their own child, and there are times when it leads to unrest. Take Ito. Whereas other users don't care about Paro at all, Ito's dementia is a bit advanced, so on the contrary, she says, "It's my baby."[16] So if it's not next to her, I feel like she gets uneasy. So on the contrary, I think it probably doesn't suit people with advanced dementia. It preys on her mind and she can't sleep or can't eat. Even when the meal is prepared, she says, "This baby is here, so first I want to see [it]." For the time being, we've put a soft toy there, but she would

rather see Paro. [Using] doll-type robots with people with advanced dementia is probably a bit difficult, I think.

Yet, these views were nuanced by an acknowledgment that Paro was acting as a form of "stimulation" for Ito. A different care worker told me, "I think Paro is really good for Ito, but she picks it up and takes it away. And when she does that, she only opens up to Paro. Then she ends up not talking to the other residents, not communicating—only with Paro. I worry that she'll lose touch with us—she likes it too much." Another summed up the mixed feelings shared by many of the staff: "One woman [i.e., Ito] loved it and treated it as if it were her own grandchild. Closing the curtain, she shut herself up in her room. It felt like she was wandering in the world of dementia. I think those times were certainly of pure bliss, but is that really OK or not?" Iwasaki related these issues to the spatiotemporal rhythms of Sakura:

> *I*: So since Ito has a great attachment to animals, when she saw [Paro], she probably wanted to make it very precious to her. So she got to wanting to sleep together with it, and without eating her meal thought, "I want to look after Paro," so yeah. Well, it can really calm the people it can calm, but if you don't take it back after a fixed amount of time, that person can become strange.
>
> *JW*: What do you think about that?
>
> *I*: Well, it has bad points and good points. Looking at it from the family's point of view, there's the side of them liking everything to follow the rules—[for the resident to] eat, sleep, have a bath, do exercise—those routines of daily life. But because the family is almost never here, the residents feel lonely, so I also have the feeling of wanting to let them do things they enjoy. But that balance—if the balance tips only in the direction of Paro, then the rhythm—of living and lifestyle—goes crazy. Because it's elderly people with dementia, they can't understand that there are mealtimes. If it's someone who can understand that, well, then I think they can absolutely use [Paro]. Because it's Ito, though, probably we have to fix the time—from this time to that time—because otherwise she'll never stop using it.

Paro experienced significantly less resistance from staff compared with Hug. Nevertheless, like Hug, Paro also created new problems of the same sort it was expected to solve. Although its introduction was aimed at improving communication, reducing loneliness, and regulating mood to ensure smooth running of the routine at Sakura, its actual use risked disruption to the institutional flow, evoked strong affective reactions from Ito, and led to a degree of withdrawal from the communal social world of the care home.

Paro and Everyday Ethical Negotiations

Devoid of a preexisting framework for interpreting Ito's new pattern of behavior, care staff searched for appropriate ways to contextualize it and in the process revealed some of the underlying friction between institutional and personal ethics of care. Care workers were not outright dismissive of Paro and clearly felt that it had a place in the care regime at Sakura. The chief concern that emerged was not that of the presumed deception involved in persuading older users—or allowing them to believe—that Paro was a sentient animal or baby, which has tended to preoccupy Euro-American scholars. After all, as described in chapter 3, care workers often deceived residents and engaged in joking imaginative play as an integral everyday element of their care and entertainment. Many residents with dementia struggled to understand where they were, why they were there, or what they were doing and required frequent reassurance from staff, who could represent reality to those residents as they saw fit;[17] their subjective lifeworlds were already somewhat removed from the institutional reality of Sakura perceived by care workers. Staff noted that Ito seemed unable to tell the difference between Paro and a real child or animal, but this was not the main target of criticism. Nor did the supposed "dehumanization" of care with social robots seem the primary concern of staff.

A related argument is, however, important to explore and has relevance to the case of Ito. Some critics have argued that talking to a robot carries a social stigma because it is viewed as weird and socially embarrassing. For example, Sone relates roboticist Okada Michio's vignette of seeing an old woman sitting on a park bench with a social robot at her side: "Watching her talking to her robot, he experienced guilt, pity, and a sense of embarrassment, as well as, contradictorily, a positive attitude that he was witnessing a (normal) episode in a contemporary, technologised society" (2017, 205). Sone suggests that the source of "shock and embarrassment" was the fact that the interaction took place in public rather than within the private confines of an eldercare institution and exemplified the phenomenon of muen shakai (a society with no ties). This was the very phenomenon that such robots were supposed to be remediating, yet in practice served to coconstruct by seeming publicly to highlight, and indeed perpetuate, social isolation. Even within the institution, care staff at Sakura articulated some of these same feelings toward the interactions between Ito and Paro. Yet, other residents and care workers expressed little negative judgment or stigma, perhaps because those residents who embraced Paro the most, such as Ito, were already substantially removed from a sense of shared reality.

In fact, the main concern highlighted by care staff was Ito's further withdrawal into the "world of dementia," conceived as a different plane of being, outside of the social and institutional space and time of the care home, which Paro appeared to encourage. This affective state of spatiotemporal withdrawal seemed related to that described by Roquet and Plourde in accounts of iyashi. Paro acted to draw Ito out of the institutional time-space and help her retreat into an escapist, subjective time-space of the self, potentially making human-human communication even more challenging. Yet, whereas Roquet argues that iyashi involves distance from troubling or upsetting affect, this seemed not entirely to be the case for Ito, who frequently cried when holding Paro and became anxious when it was taken away. When this occurred, staff expressed unease at this release of emotion that could not be controlled institutionally.

The lonely image of Ito "wandering in the world of dementia" recalled the other common type of anomalous, disruptive spatiotemporal practice of "wandering"—parsed as risky and bad because it occurred at the wrong time (at night) and in the wrong place (out of bed). It was therefore somewhat ironic that Paro, a robot intended in part to prevent wandering, should initiate it on both a psychological and physical level, prompting Ito to withdraw from the communal space of the dining hall in a way reminiscent of the phenomenon of *hikikomori* (people who have socially withdrawn). This was expressed in institutional terms as making "the rhythm go crazy": disrupting the temporally determined events of the home, most importantly mealtimes. Ito's escape from the everyday through her use of Paro undermined the institutional routine of Sakura.

Yet, the care staff were ambivalent and expressed general support for the use of Paro in spite of the problems its introduction precipitated. I attribute this to the tension between the institutional routine and the ethics and ideology of personal care and yoyū (surplus capacity). Paro in a sense brought to the surface already ongoing ethical shifts and renegotiations in care at Sakura. In reconfiguring personal time-space, Paro opened up new possibilities of and capacities (yoyū) for intimate personal care—caring for people as individuals, as humans—while in the process disrupting the fixed institutional routine. The remedy suggested by Iwasaki was to discipline Paro by bringing it into the institutional flow and setting allotted times and spaces for its use to ensure minimal disruption to the routines of the care home. This was, however, a difficult balancing act that required further supervision from staff, belying Paro's apparent ease of use.

6

PEPPER
Reconfiguring Recreation

Every afternoon at Sakura, care staff led an hour or two of recreation for residents. This involved a variety of activities depending on the preference of the day shift leader: watching a movie, singing karaoke, doing some stretches, playing a game together, or just having a chat. Like lifting, recreation is an everyday activity of care practiced in nursing care homes across Japan, and it too has become a target for automation by robotics engineers.

On one spring day in April 2017, Mr. K watched as the humanoid robot Pepper led an exercise session with a dozen residents. It was only half-jokingly that he later remarked, "I think it's better than the staff!" He added, "This would be beneficial with foreign staff" and explained that Pepper would be able to "train" new migrant care workers he planned to hire from the Philippines, who had a low level of fluency in Japanese, by modeling recreational exercise moves. They would copy Pepper's movements, and residents would copy the care workers' movements. As he put it, "We can just tell them, 'Just do what the robot does.'"

The survival of the Japanese care system has often been framed in media, academic, and state discourses in terms of a choice between the two alternatives of significantly increased immigration, particularly from Southeast Asia and China, or vastly expanded use of care robots. The unspoken assumption of the idea that robots will "save" Japan is that they will do so not just by providing care and bolstering other industries suffering from a shortage of labor but also by preventing the need for migrant workers and preserving Japan's presumed ethnic and sociocultural homogeneity. We have seen that the substitution of human care labor is a desired outcome of the Japanese government's care robotics strategy. But can robots actually take the place of migrant human caregivers and thus satisfy what are sometimes assumed by Japanese and Euro-American commentators to be the xenophobic yet technophilic care demands of the aging Japanese population? Can robots, in practice, automate care?

Robots versus Migrants

In her book on humanoid robots in Japan, Jennifer Robertson uses the heading "Robots vs Immigrants" for a section discussing robot rights discourses, contrasting the preferential civil rights treatment of robots, dolls, and even cartoon characters in Japan over the previous two decades with the more negative treatment of immigrants. She describes how "robots [are] imagined to replace the need for immigrants and migrant workers" and thus continue Japan's post–World War II approach of choosing work automation over replacement migration. She goes on,

> Humanoid robots were regarded by the public as preferable to foreign laborers, especially caregivers, ostensibly for the reason that unlike migrant and minority workers, robots have neither cultural differences nor unresolved historical (or wartime) memories to contend with, as is the case with East Asians.... Robots are perceived by some Japanese social commentators as mitigating the sociocultural anxieties provoked by foreigners. Limiting the number of nonnationals also reinforces the tenacious ideology of ethnic homogeneity. (Robertson 2018, 121–23 and 19)

One objection of Japanese nursing organizations to the Ministry of Economy, Trade, and Industry (METI)–led economic partnership agreements (EPAs) that allowed greater numbers of Southeast Asian caregivers to enter Japan was that "its provisions placed foreigners in a setting where they were working directly on Japanese bodies rather than on machines, like the majority of migrant workers in Japan had done so far" (Świtek 2016, 269–67). Tying this assumption of xenophobia to the rationale for developing care robots have been opinion polls that indicated a relatively more favorable impression of robots compared with foreign care workers among older adults, although such surveys can, as Robertson points out, "serve to reinforce and compel conformity to official views, as in this case" (2018, 19).[1]

According to Robertson, government support for the development of robots and other digital technologies represents an attempt to seal Japan off from the rest of the world and turn it into what she describes as a "technologically closed country" (*gijutsuteki sakoku*), alluding to the isolationist policy of the Tokugawa shogunate that largely restricted foreign access to Japan from the 1630s until

1854. State and industry visions of the future that emphasize robots rather than migrants, such as *Innovation 25* and Robot Town Sagami mentioned in the introduction, serve to construct and normalize an oppositional discourse between the two. At the same time, the fact that government officials, robotics engineers, robot company managers, academics, and journalists all point to predicted worker shortages as *the* basic rationale for the development of care robots clearly implies a logic of substitution for human labor. Japanese research and development of robots has been shaped by ethnonationalist politics just as much as by economic arguments.

Other imagined futures of Japan do countenance the presence of immigrants and migrant workers, and government policies themselves have been shifting. Former senior immigration official Sakanaka Hidenori (2007) has suggested a choice between alternative Japanese futures: a "Small Option" involving a shrunken population continuing to enforce strict immigration controls and a "Big Option" featuring significantly increased levels of immigration and a stable population size similar to what it was when he was writing in 2007. In advocating for his Big Option, Sakanaka was skeptical of the role of robots in enabling the Small Option. Moreover, as set out in chapter 1, despite the ongoing lack of a coherent immigration strategy, in recent years the government has significantly deregulated migration channels, with a particular effort to encourage migrants to work in care, although these policies have not yet come close to achieving their targets and it is not clear what the government's long-term plan is.

The binary of robots versus migrants is powerful because it works on several levels. It aligns with the putative and powerful (but misleading) opposition between warm human care and cold mechanical care, which makes it appealing and immediately comprehensible particularly to Euro-American commentators and their audiences. On an ideological level, a pro-(im)migration policy appears to reflect a globalist outlook, "opening up" Japan; a pro-robot policy seems to represent the techno-nationalist and xenophobic ideology of a technologically closed country. In macroeconomic terms, the two sides also seem to align neatly with alternative labor-intensive and capital-intensive strategies. As powerful anti-globalist yet techno-solutionist political movements particularly in the United States and United Kingdom, aimed in part at reducing the immigrant labor that currently undergirds their economies, have grown in recent years, the robotic replacement of service jobs in industries such as care that employ many migrant workers may appear increasingly attractive (see, e.g., Strauss 2016), and many are looking to the case of Japan to determine whether this might be economically and technologically feasible.

Robots and migrants may be frequently represented as alternative options for the future of Japanese care. But how does this framing fare in practice?

"The Japanese Staff Are Just Not There"

For several years, Mr. K had been facing the same problem as managers of other nursing care facilities across Japan and, indeed, many other countries. Care staff were in short supply, and recruitment and retention were constant challenges. He told me that he used referral agencies that typically charged one-third of a care worker's annual salary as an introduction fee, a practice he described as "outrageous." He gave the example of a sixty-one-year-old Japanese woman who had recently been introduced to him. She had no formal qualifications and only four years' experience caring for a relative. Nevertheless, the introducing agency demanded a 35 percent commission for this referral. In telling this story, Mr. K mimicked walking out the door when he heard this demand. He said there was intense competition among care homes for staff and added, apparently without hyperbole, that if one of his staff quit that day, they could stop at any of the three other care facilities on the one-and-a-half-mile trip to the local train station and be hired on the spot.

Mr. K's views on bringing in foreign care workers to address the problem of labor shortages were complex. On the one hand, he often expressed antipathy toward what he saw as the risks of unrestricted immigration. He argued that a relaxation of immigration laws had already "destroyed" South Korea, citing a recent visit there during which he had seen many Turkish and Southeast Asian migrants working in Korean businesses. He said that loosening Japanese immigration policies would soon "destroy" Japan as well, introducing an influx of foreign workers into the service industry who would probably not learn to speak Japanese fluently or fully respect or integrate into Japanese society and culture. In particular, he argued that if the proposed expansion of the government's guest worker program, the Technical Intern Training Program (TITP), went ahead (as it did in 2017), over the subsequent years, "They're gonna lose it all; they'll be taking the lid off the jar. . . . If this happens, Japan is finished."

Yet, Mr. K's views were more nuanced than this statement might suggest. At times, he told me that he hoped the expansion of TITP *would* go ahead. As he put it, "The day is coming—and it's coming very quickly—it's *here*—where the Japanese staff are just not there." He had hired several Southeast Asian care workers who were already permanent residents in Japan and had provided them with twice-monthly Japanese classes held at Sakura during paid working hours, with cash bonuses if they passed Japanese Language Proficiency Test exams. As permanent residents, they were not subject to the restrictions of any government migration program.

Mr. K had also tried to participate in the government's EPA scheme to hire care staff from the Philippines, Indonesia, and Vietnam. He told me that EPA was expensive because care home managers were legally required to pay EPA workers the same as Japanese workers, while having to deal with additional bureaucratic hurdles. Participating in the scheme involved hiring someone without being able to interview or vet them, at the same cost as a Japanese care worker. It was highly unlikely that this person could speak Japanese very well, and they would therefore not be able to do a full job for the first couple of years, if at all. The main obstacle, however, was the fact that EPA care workers were only permitted to stay in Japan for four years before having to pass the formal qualification exam to become a certified care worker or face returning to their home country. He described this requirement as "absurd"—particularly since fewer than one-third of Japanese care workers themselves held this qualification. His experience with the EPA scheme had not been satisfactory and in the end, he withdrew his participation. After an introductory matching event for employers and prospective employees, facilitated by the government officials operating the program, he had specified which of the candidates he had met that he wanted to hire. The officials in charge, however, told him he was not allowed to select individual care workers—companies were only allowed to specify how many workers they wanted, and the officials would simply assign them that number of people. This seemed to indicate that for the officials, care workers were interchangeable and care companies should accept whatever labor they were given. Indeed, several large care companies had accepted these conditions and recruited through the scheme; these larger corporations appeared to be the main targets of the program.

Mr. K's approach toward migrant care workers revealed complex and sometimes contradictory attitudes toward foreigners that were echoed in interviews with Japanese staff members at Sakura. Although some told me that nationality did not matter in the slightest to the residents and others noted that foreigners could introduce fresh ideas and often worked hard, most said they thought that nationality *did* make a difference—primarily due to problems of communication. For example, Otsuka said,

> Honestly, I don't think [employing foreign care staff is] good. Of course there are communication problems, and even for us [i.e., Japanese care workers], mutual understanding with residents is quite hard.

One of the floor managers explained, "If we use [foreign care workers] in some sense as manual laborers, I guess we can use them. Right now, Diego [a Japanese Peruvian care worker] on the second floor is doing care, and as long as they can communicate with words, then I think it's OK. But there are all kinds of other things involved—reading reports, keeping records, and so on." Murata described foreign care workers as "less detailed than Japanese people," and other

Japanese staff talked about the impact of a different cultural upbringing on values and practices. For example, Iwasaki explained,

> Because they didn't grow up in Japan, in terms of Japan's culture, well you probably can't beat a Japanese person. In terms of work or language and communication. They try their best at the job, but in Japan everything has to be perfect, so in that sense it's a bit different.

A senior member of staff said, "I think they're not bad, but, well, if you ask, 'Is there a difference from a Japanese person?' well, I think there's definitely a difference. If you ask which one [is better at care], then of course a Japanese person is better. More serious." Another care worker summed up some of these attitudes: "When I was young, there weren't foreign people working, so I halfways think it's interesting to try here, but when it comes to things like not being able to communicate, not being able to tell subtle nuances, and so on, then I think, ahh, yeah, it's difficult! [laughs]" Although these Japanese care workers held ambivalent attitudes toward foreign staff, the main limitations they identified were communication and cultural difference.

Despite Mr. K's unfavorable views toward certain types of guest workers and migrants, his actions reflected a pragmatism that has led many employers in the care sector to try to make the most of the flawed EPA system as well as the other more recently opened migration channels, or to find other workarounds to cope with domestic labor shortages. Nevertheless, the idea of using robots to supplement and ultimately substitute for human care workers and avoid the costs, risks, and complexities of hiring and training both migrant and local care staff was clearly extremely attractive. In late 2016, while we were planning which robots to trial at Sakura, Mr. K emailed me:

> I DON'T want to spend all that money for a talking wiggly doll. I want a REAL robot, roaming the floor, actively interacting with residents. Even playing preprogrammed music for specific residents if possible. A real robot.

Out of the three robots trialed at Sakura, the one that seemed to conform most closely to his expectations of a "real robot" was Pepper.

Pepper

Pepper is a four-foot-high humanoid robot with abstracted and cartoonlike facial and bodily features and an outer layer made of white plastic (see figure 5). An article in the *New Yorker* memorably described it as looking like "a cross be-

FIGURE 5. Pepper unboxed. Photo by author.

tween a mermaid and the Pillsbury Doughboy" (Marx 2018). It has a touchpad mounted on its chest, large eyes that light up in different colors, and round circles for ears. The head contains both microphones and speakers to enable Pepper to communicate with users, as well as cameras for vision and touch sensors to detect when it is being patted on the head. It has arms, hands, and rubbery tactile fingers and a wheeled podium in place of legs; the head, arms, and hands can move, and it can also move around on its wheelbase, enabling it to perform dance moves or model upper body exercises. By default, it speaks in Japanese with a rather shrill, high-pitched voice; its English voice, intended for use in North America and Europe, is deeper. Pepper has no officially designated gender and is programmed to speak and behave in a manner that is childlike, cute, somewhat naive but also intended to seem emotionally intelligent and humorous. Its sleek, shiny white design is evocative of devices like Apple's iPhone, with which Pepper shares the same Taiwanese manufacturer, Foxconn. Pepper serves as a hardware platform for its apps, which are mostly provided by external developers, and are accessed from the touchscreen on its chest. Like most care robots on the market, Pepper was not specifically designed for eldercare but was repurposed for care home recreation via specialized apps.

Pepper was perhaps the most publicly visible embodiment of Japanese tech corporation SoftBank's ambition to become the world leader in service robotics and other cutting-edge technologies. SoftBank acquired Aldebaran, the French company that designed Pepper, for $100 million in 2012, rebranding it SoftBank Robotics. By 2015, the year that Pepper, its main product, was launched in Japan, SoftBank Robotics was valued at $590 million. In 2017, SoftBank also acquired major robot companies Boston Dynamics and Schaft from Alphabet (Google's parent company)—part of a far larger strategy to invest in pioneering technology via its gargantuan $100 billion Vision Fund.² Pepper thus constituted not only an international investment vehicle potentially worth hundreds of millions of dollars but also a symbol of Japan's leadership in the emerging market of consumer-facing service robotics, while embodying the promise of "robot revolution" domestically. In the couple of years leading up to my fieldwork, national and global corporate expectations for Pepper were extremely high, and Pepper represented the most extravagantly hyped, highly capitalized, and widely marketed "social" service robot in the world.

Pepper's "personality" had been as carefully engineered as its glossy appearance. It was unveiled at a major launch event during which, in a dramatic and meticulously choreographed performance, SoftBank CEO Son Masayoshi symbolically handed Pepper a red heart. Pepper was promoted as the "world's first humanoid robot that reads human emotions" and branded as "an emotional companion."³ As part of an advertising campaign in Japan, a SoftBank advert began with the caption "Just a little in the future" and featured a vignette of Pepper comforting a crying woman. After detecting that she is upset, Pepper tells a joke and then displays a message from a man we assume to be her partner, apologizing for his bad behavior; the woman, overcome with relief, embraces Pepper, as if it were embodying her partner at that moment. Apparently confused, Pepper asks, "Even though you're OK, you're crying? I guess I really don't understand women's hearts."[4] Another television advert claimed that Pepper was "the world's first robot to approach people,"[5] employing the same spatial metaphor for reducing distance used by care staff at Sakura, with its ethical connotations of affective proximity.

At the time of my research, Pepper was being marketed as a multipurpose service robot that could attract and emotionally engage with customers and other users, and provide information or services to them while, importantly, collecting data from and about them. SoftBank had leased twenty thousand Pepper units by April 2017, around the time that it was trialed at Sakura, although only around five hundred were being used in nursing homes as of early 2018 (Liu 2017; Foster 2018).[6] Pepper would have cost Sakura approximately ¥2.5 million ($25,000), leased over a period of three years. This price included a small number of apps designed for care homes, which each cost around ¥10,000 ($100) per month for subscription. Even with state subsidies to cover some of the cost, Pepper would have represented a very significant investment for a publicly funded care home. SoftBank Robotics agreed to lend a Pepper model to Sakura for a free trial lasting six weeks.[7]

Before Pepper was introduced at Sakura, Mr. K said he imagined the robot greeting residents in the morning and directing them to the dining hall for breakfast, conducting exercises and quizzes during recreation time, and even delivering whole presentations to senior staff in the care home. He told me that if, as SoftBank Robotics claimed, Pepper's exercise apps were designed by fitness professionals, then "Pepper would have a higher level of qualification than any of my staff." Yet, many of these expectations—heightened by the way the robot had been marketed—were not ultimately met. In the end, Mr. K and the staff decided to use Pepper primarily for recreation sessions, its most common use in Japanese care facilities at that time.

During my fieldwork, I met and interviewed a number of people involved in various aspects of the emergent "Pepper economy" of apps, accessories, products, and services related to Pepper, at SoftBank's headquarters, Atelier Akihabara (a drop-in training center for Pepper users and programmers), Pepper World (a regular exhibition event featuring new apps and accessories), and businesses already using Pepper. Many of them revealed stark differences between the hype surrounding Pepper and its actual functionality. Despite the effort put into establishing the empathic image of Pepper as an emotional companion, in real-life switching on its "emotion-reading" functionality, based on interpreting the user's facial expressions, involved a complex technical process and played no role in its use at Sakura—or indeed, almost any of its other commercial applications. SoftBank Robotics' marketing also fed a myth of autonomy, constructing an image of Pepper as a stand-alone robot with its own personality that could simply be left (in Mr. K's words) "roaming the floor, actively interacting with residents," potentially substituting for human labor. The staff at Atelier Akihabara, however, told me that there always needed to be someone assisting Pepper to do things so that it could only save labor for activities that already required several staff members, reducing labor requirements incrementally, if at all.

Promotional videos showcasing Pepper's use in care homes explicitly tried to demonstrate how the robot could replace care workers during specific activities such as recreation and thus provide a return on investment. But in these videos, as was the case in several optimistic Japanese TV documentaries and news clips about the use of care robots in institutional environments at the time, there were always plenty of care workers on hand—in one video posted by SoftBank Robotics, five care workers were shown overseeing the use of Pepper with five older care home residents. Of course, this ratio did not reflect the reality of Sakura, where, as in most care homes in Japan, staff were so thinly spread that usually only one person would be available to do recreational activities or exercises with residents. In another video, Pepper was portrayed autonomously patrolling corridors at night to deal with older people with dementia wandering around. These and other corporate depictions of Pepper constructed an imagined institutional care, peopled by imagined care staff and older adults, in which Pepper would act effectively in imagined ways that (by SoftBank contract staff's own admission) misrepresented its actual functionality.[8] It therefore seemed unlikely that Pepper would act as a labor-saving device at the genba (actual site) of Sakura; nevertheless, there was an air of excitement among staff at the prospect of having a "real robot" in the home.

Pepper in Practice

Pepper was delivered to Sakura in a six-foot-high cardboard box covered with lavish illustrations of the robot inside. The interior of the box was decorated with a night sky filled with stars and phrases such as, "Is it true the sky is blue?" written in English, as if Pepper were asleep inside, dreaming of the outside world. The packaging was playfully designed to encourage the user to anthropomorphize Pepper and, by extension, to inspire an affective and indeed caring relationship toward it from the moment of unboxing (see, e.g., White 2018). Later that day, two managers from SoftBank Robotics visited to set up Pepper's software. After three hours of trying and failing to get Pepper to work, they announced that the unit was not functioning correctly and would have to be replaced. This was the first of a series of technical problems that occasionally beset Pepper. The robot took variable amounts of time to set up, sometimes requiring a reboot and sometimes simply freezing. The Wi-Fi connection, emanating from a mobile broadband device supplied by SoftBank, came and went, intermittently rendering apps that required an internet connection inoperable and creating uncertainty as to whether Pepper would perform correctly on a given occasion.

Care workers were initially invited to a short learning session where the two SoftBank representatives explained the basics of how to operate Pepper and its care apps. After discussions with the floor managers, Mr. K and I asked staff to try using Pepper during recreation time in the afternoon and at any other times they saw fit. Several apps were available, including Ritsuko's Rexercise (a portmanteau of "recreational exercise") and Everyday Robo Recreation—both of which combined group exercises and singing.[9] Other apps could make Pepper sing and dance in preprogrammed routines, tell jokes, and make conversation, although staff used these less often during the trial. This was partly because, as described in the vignette at the start of this book, the background noise at Sakura rendered Pepper's speech recognition function largely ineffective.

As one of the few occasions during the day when care staff could take time to communicate with residents without having to rush to do other tasks, recreation was an extremely important part of care at Sakura; several care workers told me it was their favorite part of the job. Activities were varied; physical exercises always constituted one part of recreation but usually only lasted for a few minutes. By contrast, rexercise sessions with Pepper (see figure 6) were longer and followed a set format. The shift leader first retrieved Pepper from where it was stored in the office. They checked that it was charged and "woke it up" from sleep mode, rebooting if necessary, before wheeling it over to the front of the room and reintroducing it to residents. The shift leader used the touchscreen to select the rexercise app and the length of the program (either twenty or thirty minutes). Pepper would then start its programmed routine, first playing some upbeat music and pepping up the audience before launching into gentle stretches that gradually became more challenging. Throughout the session, Pepper called out instructions in its high-pitched voice, and the shift leader, standing next to it, echoed them and copied Pepper's motions. At the end of each section of the routine, Pepper thanked residents for taking part and asked someone to pat its head. The shift leader would help a resident come to the front and pat Pepper, which reacted with one of several humorous responses. The next section involved Pepper leading residents to sing a traditional Japanese children's song while making accompanying movements and stretches. After another pat on the head, Pepper proceeded to the final part of the routine—more exercises and another song—before ending with several *banzai* cheers, robot and residents saluting each other with ten thousand years of long life. Pepper told jokes throughout its routine and used cute, childlike onomatopoeic diction that seemed to delight residents and staff. Most residents copied Pepper's and the shift leader's movements, even continuing to mimic them after the routine had finished. Pepper played some relaxing music to

FIGURE 6. Pepper and a care worker conducting "rexercise." Photo by author.

wind down and asked, "How did you feel today?" Some residents replied directly, thanking Pepper for its efforts.

A key difference between the recreational sessions run by care workers before Pepper's introduction and the rexercise provided by Pepper was the fixed timing and standardized routine of Pepper's program, compared with the variable timing and activities led by human care staff. With Pepper, the staff selected the length of the program and, depending on the app, the songs that would be played. Then the robot simply carried out the program. The workouts provided by Pepper were longer and more elaborate than those run previously by human staff, and residents appeared engaged and happy with them as long as a care worker stood next to Pepper, repeating Pepper's words and mimicking Pepper's movements. When staff did not participate, there was significantly less interest or participation from residents. Following its relatively positive reception, Pepper began to be used almost every afternoon. Although it came to be used less frequently over time, staff continued to wheel it out every two or three days until the end of the six-week trial, when it was packed up and returned to SoftBank Robotics.

Several care workers who had said they were bad at running recreation activities told me that Pepper was very good at helping them mediate recreation and made it more fun. Pepper also helped fill in for the shift leader during brief interruptions. Kubo explained how

> during recreation time, only the shift leader is left on the floor. You can't just leave it to a machine—if [the residents] just touch it or something, it could break. But when you have to stop doing recreation when you're

called away for a moment, because you're on your own, I thought it was really good at that time that it could keep talking.

Other comments were more negative, such as this from a male care worker in his thirties:

> People who are hard of hearing can't hear Pepper's voice. People who are sitting down can't see it. There were people who weren't interested, or who said they were scared. During exercise time, since those at the back couldn't see, a staff member stood next to Pepper and did it together.

There also seemed to be subtle tensions about money. No member of staff complained to me directly about their salaries. But one care worker, when asked his thoughts regarding the robots, answered that the staff should receive higher wages: if the home had money to spend on expensive robots like Pepper, then it should be able to afford to pay the staff more. During my time at Sakura, the most frequent question staff asked about the robots was how much they cost, and many expressed shock when told the prices—perhaps an indirect way to critique the logic of considering purchasing such expensive machines.

A striking characteristic of staff descriptions of Pepper was the similarity to the language they used when talking about residents. They pointed out that Pepper needed staff assistance, both in terms of time and physical effort, to move around. As with residents, it was at risk of toppling over, which in turn could lead to it "injuring" itself. Similarly, it required staff to monitor it (*mimamori*; literally meaning "watch and protect") in case it fell over or something went wrong, to make sure that it was interacting suitably with residents, and to ensure that residents would not trip over it. Mimamori had been one of the very functionalities of Pepper marketed by SoftBank, although the mimamori app it promoted had not been used at Sakura because it did not work effectively in real life. Rather than Mr. K's imagined use of Pepper to watch over residents, it was Pepper that needed to be watched over by care workers. As its speech recognition system struggled and, ultimately, failed to cope with the background noise at Sakura and it sometimes took several seconds to process things it was told and formulate a response, it was described by care staff as "hard of hearing." Pepper's inappropriate or mistimed responses became a source of humor for staff and residents, with its vulnerability transformed into cuteness in a similar way to how staff described some residents. Over time, as staff became habituated to its presence, Pepper seemed to "settle down" into its existence at Sakura rather like one of the human residents. In the end, it was not realistic that Pepper could act in a stand-alone way, roaming the halls and actively interacting with residents as Mr. K had envisaged. Like residents themselves, Pepper was dependent on the care of staff.

In fact, the use of all three care robots at Sakura was predicated on additional human labor.[10] The 143-pound Hug lifting robot needed to be pushed slowly and carefully from room to room, stored and recharged, and took more time to operate than simply manually lifting residents. It was partly because of the additional time and effort required and the reduction of yoyū—the "surplus capacity" for personalized care that could be spent with individual residents while still completing the institutionally set tasks—that staff had quickly rejected the use of Hug. Even Paro, seemingly the most hands-off robot given its limited functionality and intuitive usability, required careful observation by care workers, particularly in the case of Ito, the resident who developed the strongest attachment to it.

In the years following the end of my fieldwork, I continued to stay in touch with Mr. K to find out how the situation at Sakura had changed since the care robot trial. In 2018, he told me of a new plan for employing foreign care workers. Instead of recruiting qualified care staff under the EPA scheme, he had started hiring foreign students from Mexico and the Philippines. Entering Japan on "long-term care" study visas, these students could legally work part time at Sakura—although "part time" was defined as up to twenty-eight hours per week during term time and forty hours per week during school vacation. This plan did not work out, however: the students had wanted to move into a better paid job in a different industry at the end of their studies.[11] By 2020, Mr. K had switched to recruiting through the TITP program, which had become, he said, easier to access with more private recruitment companies operating in the market, although the arrival of new workers had been delayed due to the COVID-19 pandemic. He said his long-term aim was eventually to phase out his Japanese staff altogether as they quit or retired. Although the only new robots he had purchased since the end of the trial were robotic vacuum cleaners, he was continuing to trial new devices and told me that he expected robots like Pepper or its successors to become increasingly useful for "training" his incoming foreign staff and helping them to adjust to Japanese care work.

Deskilling, Distance, Displacement: Toward a "Culturally Odorless" Care?

In Japan, robots were intended by politicians and engineers to save time and money by substituting for human care workers and thus reducing the need for migrants, particularly from Southeast Asia. As philosopher Jennifer Parks notes, robotic care technologies are commonly believed to "enable some of the most time-consuming and labor-intensive care demands to be addressed so that human caretakers may spend their time on other care tasks," although she warns that

> economic pressures make it highly unlikely that nursing homes will maintain the same number of human staff members once much of the work can be delegated to machines and robots. The likely consequence of a technology boom in aged care is that the number of human caretakers will be seriously reduced; the net result will be a further reduction in the amount of human contact to which our elderly citizens will have access. (2010, 116)[12]

But in reality, such robots currently place even greater demands on the limited time available to staff in a typical midsize institution such as Sakura as well as more cost pressure on cash-strapped publicly funded care homes. Despite the claims of SoftBank Robotics, and the expectations of Mr. K, Pepper did not reduce the need for human labor at Sakura. In fact, if all of the test robots had been used every day, Mr. K would have required more human care workers.

Technologies "do not work or fail in and of themselves. Rather, they depend on care work" (Mol, Moser, and Pols 2010, 14). So it was with Pepper and the other robots: far from being autonomous, they relied on care workers, requiring additional time and effort from staff. This extra human labor was hidden in plain sight, discounted in promotional videos and overlooked in enthusiastic state strategy documents but keenly felt by care staff who were sensitive to any change in the flow of daily life because of the tight constraints on their time.[13]

The introduction of new technologies does not necessarily result in the deskilling of workers. Some innovations may involve not only deskilling but also upskilling, sometimes eliminating repetitive and boring duties and facilitating more "creative" or complex tasks, with the net impact for workers often ambiguous and difficult to quantify. Just as Pepper did not entail a precise substitution of capital for labor but instead introduced new relations of labor *with* capital, nor did it involve a straightforward deskilling of care. Rather, the skills required changed, as recreation became a site for virtually assembling skilled digital labor physically dispersed across research institutes and companies involved in the development of Pepper and its apps. The expansion and fragmentation[14] of the division of care labor that Pepper enabled and indeed required, and the "Pepper economy" that was mobilized around it, in turn involved a reassignment of value, diminishing the value of intimate human social care while privileging the more abstract and highly paid skilled labor of software developers and robotics engineers and bringing their algorithmic vision of care into the home. For care workers at Sakura, the potential for the deskilling and devaluation of their labor was immediately apparent. Whereas previously, care staff often planned or improvised a variety of recreational activities based on their knowledge of exercise techniques as well as their social skills and relationships built up over time with individual residents, they now only had to wheel Pepper out and mimic its standardized physical actions and routines in set blocks of twenty or thirty minutes. Linguistic and communicative social skills used during recreation could be replaced with the modeling of bodily movements or, as Mr. K put it, "just do[ing] what the robot does."

Just as Pepper eliminated the need for fluency in Japanese and a knowledge of Japanese culture as prerequisites for conducting recreation sessions, Paro, as a cute, tactile medium for communication that could keep residents occupied, eliminated some of the need for direct verbal interaction between care staff and residents, which has often been seen as a fundamental barrier for foreign care workers in Japan. Hug, in turn, eliminated much of the need for foreign care staff to directly touch older Japanese bodies during lifting, which has likewise been identified as a reason for rejecting migrant care. Other devices included in broad state definitions of care robots, such as high-tech portable toilets and bathing machines, could reduce tactile interaction even further.

Iwabuchi Koichi has described Japanese consumer electronics as "culturally odorless," stripped of their national cultural specificity for foreign consumers (2002, 24). Odor is a prominent sensory characteristic of the care home; negative preconceptions about residential care frequently center on the smell of excrement and decay. In fact, the first impression upon entering a care home is usually the pungent chemical smell of detergent. Odor is also a productive metaphor in the context of care, implying a familiarity, a closeness, and a culturally mediated whiff of pollution; as Anna Tsing writes, "Smell is the presence of another in ourselves" (2015, 45). In reconfiguring care practices and skill, robots also recalibrated the distance, both physical and metaphorical, between caregivers and care recipients. Yet by shifting care away from intimate, immediate, and human care toward the abstracted care of programmers and engineers and by introducing new physical and social distances between bodies, care robots such as Hug, Paro, and Pepper have the potential to render interactions between care workers and older Japanese care recipients culturally odorless, bypassing the "problems" of culture and communication that were identified by Japanese care workers at Sakura as the main obstacles to using foreign staff. People, like robots, can thus become standardizable and interchangeable units free from the perceived constraints of social or cultural relations—a result that seems to extend, rather than remedy, the issue of muen shakai, or a society free of meaningful personal ties.

Instead of a binary opposition of care robots and foreign care workers, which has been presumed to be a primary driver of care robot development, the relationship between robots and foreign care staff is more symbiotic than oppositional. Mr. K told me that he foresaw a near future in which robots like Pepper would be combined with migrant care workers to deliver a more "skilled" service (such as rexercise) than could previously have been done by Japanese care workers. At the same time, the fact that Pepper could sing old Japanese songs with which residents were familiar, and which foreign care staff might not know or might struggle to learn, would provide peace of mind to residents to accept both robots and foreigners. Rather than eliminate the need for foreign caregivers, widespread use of Pepper and other robotic devices could both necessitate and facilitate large increases in migrant labor, serving as fulcrums for a new, more globalized assemblage of techno-care. In this way, Pepper in practice represents the precise opposite of the ideology of a "technologically closed country" described by Robertson.

Preventing the need for care workers to hand-wash or manually lift aged bodies may appear socially and ethically unproblematic and indeed highly desirable—not least at a time of accelerating pandemics. Perhaps robots will make it easier for foreign workers to deal with the often-intimidating experience of working in Japan, brokering relationships with care recipients, and reducing the friction in care as sociocultural practice—reflecting back an image of sleek Japanese modernity that could make migrant care more acceptable to service users. In implementing new technologically facilitated care practices, however, it is important to consider what is lost as well as what is gained: what other skills and techniques risk being replaced in the process of roboticization.

Many care workers told me that recreation was one of the most enjoyable parts of their job and a central element of personal care, affording them time to communicate with residents both en masse and one-on-one. Pepper's envisaged use suggests that it could reduce the need for human-human verbal interaction in the future, alienating staff and turning more components of care work into straightforward manual labor. In this sense, Pepper shares some similarities with Hug, whose use likewise reconfigured the task of lifting to depend less on cues of bodily contact, tactile perception, and communication, instead involving operating machinery. In the cases of all three robots, care duties that previously had involved skilled social interactions of tactile and joking care and rapport—duties that were suffused with an ideology and motivational ethics of personal care—could have this human-human interaction stripped away, with the robot taking center stage and human care staff themselves becoming more "robotic." By reducing sources of intrinsic motivation, greater use of such robots could make care jobs even less attractive to Japanese care workers and further shift the moral economy of care work.

Moreover, by removing skill and training barriers and by repositioning where value lies in care work, the introduction of a new robotic division of labor is also likely to affect care staff's extrinsic motivation in the form of future pay, facilitating the further commodification of care and aligning with ongoing neoliberal processes of precaritization in Japan (Allison 2013) and perhaps beyond. Care robots seem to offer standardization and scalability as well as the prospect of rendering human care workers themselves more interchangeable, although any return on investment only appears feasible for operators of larger care home groups that can better leverage investment capital, make use of economies of scale, and successfully navigate government migrant worker schemes and subsidies to purchase robots.

Bearing out some of these findings, a report by the U.S. National Bureau of Economic Research that analyzed economic data on the use of care robots in Japan found that care homes that adopted one or more care robot employed more people than homes with no robots but that this difference was entirely accounted for by nonregular employees (Eggleston, Lee, and Iizuka 2021). At the same time, "robot adoption reduces monthly wages of nurses, especially of regular nurses" (17). The report also finds that homes with robots were more likely to have hired foreign care workers and to have active plans to hire more in the near future.[15]

The new assemblage of techno-care that the use of such robots presages, involving more globalized, culturally odorless care, is ironically premised on maintaining a veneer of national cultural acceptability in the form of Japanese songs and discourses of Japanese craftsmanship, while at the same time relying on a muted, highly precarious, and socioeconomically marginalized migrant workforce. Residents and workers alike may no longer be able to form the kinds of close, sustaining social relationships so critical to care at Sakura. Relying on robots as cultural translation devices may mean that no one has to make an effort at mutual understanding.

Regardless of the intentions of developers, technocrats, and politicians, the objective of robotic substitution for human labor in the field of institutional care remains elusive for the time being. Instead of replacement, a better description for the relationship between care robots and human caregivers is *dis*placement. The introduction of care robots displaces skills, practices, and value; their mediation displaces direct human-human contact; caring for robots risks displacing caring for people.

7

BEYOND CARE ROBOTS

On a sunny spring afternoon, the floor manager, Maeda, closed the curtains in the dining hall and connected the TV to the DVD player. This pattern repeated itself every time Maeda was shift leader and in charge of afternoon recreation at Sakura. Self-admittedly "bad at" standing up in front of the residents and leading recreational activities, he almost always played a movie and almost always the same one. It was a movie that seemed to hold the attention of the residents, allowing him to return to the office and complete his record-keeping duties on the computer.

As he hit the play button, a black-and-white film appeared on the screen: the popular Charlie Chaplin comedy from 1936, *Modern Times*, in which Chaplin's Little Tramp character (referred to in the credits as "a factory worker") must navigate the hard and hectic times of American life during the Great Depression, depicted as a period of mass unemployment and industrial automation, communist marches, and union strikes.[1]

In one surreal sequence, the Tramp falls inside the machinery of the factory and is churned through a series of cogs and wheels—literally chewed up and spit out by the apparatus of industrial capitalism, providing a classic depiction of the absurdity of the human caught in the relentless and unyielding logic of the machine. The film highlights the Tramp's inability to synchronize with the temporalities of the "modern times" of rationalized and technologized institutions, which operate at breakneck speed in monotonous uniformity. This provides both a source of physical humor and a serious theme of the movie—a critique of the industrial machinery of modernity that forces humans to adjust to its own rhythm. The "modern times" portrayed in the film were novel, frightening, desperate: a time of economic and labor upheaval and exploitation apparently driven by industrial technology overseen by ruthless capitalist bosses. Amid the dehumanization of the factory, escape had to be found in humor.

Modern Times came out a decade and a half after Čapek's play *R.U.R.* introduced the concept of the robot, and the movie emerged from a very different context. But it explored a similar, still unresolved theme: the place of the human and their labor in the modern times of industrial technoscience. Čapek's notion of a robot can be understood as a metaphor, and the story can be read as an allegory about the revolutionary dangers of dehumanized work in the factory or office. Rather than mechanical androids, Čapek's robots seem more like an underclass of people stripped of their souls by capitalist exploitation and scientific management—turned back into *robota*, an old Czech word for feudal forced labor, both worker and product, labor and capital.² Eventually, the robots organize, issue a manifesto, and rebel against their masters, wiping out humanity. Although *R.U.R.* dealt with science fiction and the hubris of men using technology to play god, it was also about exploring what might happen if industrial capitalism ("progress")—were taken to its extreme. In a sense, these robots are a mirror image of *Modern Times*' Little Tramp, although the comedy, while equally absurdist, is far blacker.

Modern Times proved an apt reference point for my fieldwork. The human rhythms of time and space in everyday life and care at Sakura and the problems of asynchrony—machines operating out of sync with the lives and work of the people using them—ran through my research. The Japanese government saw robots as promising a new industrial revolution—a robot revolution—and a renewed cycle of modernity. Japan, like the rest of the world, seems to stand again on the brink of a new wave of roboticization in which machines such as care robots will increasingly be introduced into our lives, but in which we, like the Little Tramp, may find ourselves the object rather than the subject of this "revolution."

The three robots trialed at Sakura seemed to symbolize some of the promise of this revolved modernity, while their use revealed complex interrelationships between time, space, people, and machines, as they reconfigured some care activities and mediated new spatiotemporal arrangements of skill, labor, and capital in the care home and beyond. Yet despite his enthusiasm for introducing care robots and the seemingly pro-robot local and national policy environments, Mr. K decided neither to purchase nor lease any of the three robots after the trial ended. This lack of adoption is writ large across Japan. The Care Work Stabilization Center (2020) survey of over nine thousand eldercare institutions in Japan showed that in 2019, only about 10 percent reported having introduced any care robot (most of which were monitoring or communication robots), although actual usage was likely to be even lower, particularly as the same survey showed that only half of care workers where robots had been introduced reported finding them useful.³

It seems clear that care robots have not yet scaled up in the way envisaged by Japanese state and industry technocrats. What went wrong, and what can this tell us about the Japanese government's broader aspirations for the national and international roboticization of eldercare? What lessons does Japan's experience hold for the rest of the aging world?

Back to the Future

For much of the twentieth century in Japan, robots symbolized the future. *R.U.R.* was performed in Tokyo in 1924, just a year after it premiered in London, and had a significant impact, sparking a robot boom—a popular fascination with robots, a flourishing of robot-centered sci-fi stories, and the creation of early prototype humanoid robots. These works explored the potential for robots to pave the way toward an internationalist, postracial, utopian society and also imagined uncertain, potentially troubling, and morally ambivalent sci-fi futures, reflecting concerns at the time about the rapid industrialization that was transforming Japanese society (Nakamura 2007; Frumer 2018). Many of these themes continued in post–World War II manga and anime stories of robots such as Astro Boy, Mazinger Z, Gundam, and Doraemon. The factory robot boom of the 1970s and 1980s was about escaping a troubled past and propelling the country toward a prosperous industrial future of modernity. Industrial robots not only demonstrated technological mastery of space, time, and manipulation but also materialized economic mastery and nation rebuilding. As anthropologist Anne Allison notes, "Robots . . . became the tropes and fantasies of the postwar era by which Japanese crafted a new imaginary of the state and themselves" (2006, 56).

The symbolism of robots in Japanese society helps explain why the development and implementation of care robots became such a significant project for the Ministry of Economy, Trade, and Industry (METI) and the Cabinet Office, which have presented them as a way to reclaim the future and revitalize an enervated Japan by recapturing a sense of the forward momentum of history, recovering a national identity rooted in technological modernity. Robots also seemed to offer the possibility of remedying the substantial loss of public confidence and trust in both technology and the government following Japan's 2011 earthquake, tsunami, and nuclear disaster. Care robots in particular encapsulate a central paradox of contemporary Japan, representing and contributing to the aspirations of an imagined technological future of progress, but at the same time serving to reify national anxiety and even collective despair over the perceived demographic, economic, and social decay of the Japanese society and nation.

Nursing care homes serve as a focal point of concern around perceived national entropy. Ambiguously positioned between private and public, commercial and domestic, business and home, they are places for older people and their caregivers to spend time unproductively, according to the logic of the market—spatiotemporal bubbles where the national preoccupation with productivist time holds less sway and therefore also sites of friction with the neoliberal instincts and policies of the governing Liberal Democratic Party.

Temporalities of the Care Home

In this book, I have adopted the metaphor used by care workers of nagare, or the "flow" of time and space in the care home that delineated daily life and care. When viewed at the micro level, the timescape at Sakura was complex and consisted of multiple intertwining and overlapping temporalities.

The flow of institutional time was heavily routinized, segmented into blocks recorded on timesheets, and marked by accelerations and decelerations depending on the number of care staff available and the number of duties to be completed at any one time. Residents, on the other hand, had their own slower and less linear temporalities. For many residents living with dementia, time seemed circular, operating in loops, with some asking to go home at around the same time every afternoon, or making repetitious requests of staff. Time for personal care, by contrast, was imbued by staff with ethical value, and they represented it as a prerequisite of "good care." This meant personalized time with residents, treating them as humanized individuals. In other words, it required yoyū—the surplus of time-space outside of the logic governing Japanese productivist capitalism and commodified care.

Yoyū was particularly associated with slow communication practices such as recreation and chatting but was also invoked during other everyday work duties such as lifting. Creating the time-space for personal care meant stepping outside of or manipulating linear institutional time—not just doing one thing after another—in order to engage in meaningful communication, providing the opportunity and, in resistance to the work demands of the institution, the leisure, to "approach" and accumulate relationships with residents over time through repeated caring encounters. Touch and tactile care were critical to this process, enabling some relationships to become familial over time. This individual time spent together in turn provided much of the job satisfaction and motivation for care staff. In addition to referring to the care given to residents, personal care was also in a sense care for the personhood of care workers themselves, who constructed narratives about the ethical value and meaning of their labor based around it.

Despite the superficially fixed appearance of institutional flow, the overall temporal ecology at Sakura was dynamic. Care might have looked the same year in, year out, but because it was a sociotechnical assemblage with relational, material, and political-economic dimensions, it shifted over time. Introducing new residents who required more care or losing a care worker or two would have an impact on the whole environment. Spending time on one demanding resident meant losing time to deal with others. Introducing a new communication technique or technology might save time or take time; time itself was conceptualized by staff as a scarce but elastic resource that could be manipulated, made, and lost. It was also a site of resistance—part of a negotiating position for care workers who could reject management demands by saying simply, "We have no time," glossing over the temporal ebb and flow during the day and wresting more power over the kind of care they wanted to do. Time is invisible and intangible yet inherently political and entangled in arrangements of institutional power.

Care workers attempted to synchronize these competing temporalities, and personal care, in particular, was a way to harmonize the demands of institutional time flow with the subjective time experienced by the residents. The tension between institutional time and personal care time created much of the job dissatisfaction for care staff, belying the usual explanation that the job was dirty, dangerous, and difficult. Asynchrony between care workers and residents was also inscribed with risk: for example, pushing a resident's wheelchair too fast or feeding them too quickly could lead to accidents resulting in serious injury or death, while taking too long to take them out of the bath could expose them to the risk of catching a cold. This asynchrony expressed a growing contradiction between the care home as a communal institution and a newer ideology of individualized personal care playing out at nursing homes like Sakura particularly since around 2000, in tandem with the introduction of the Long-Term Care Insurance (LTCI) system.

JaneMaree Maher argues that industrialized temporality—seeing time as a finite resource to be portioned out in fixed blocks—"is understood to devalue and reduce the time for care, which does not function according to the same temporalities as paid work; care time doesn't produce 'product' in the same way that industrial time does" (2009, 234). She argues that although feminist activists in the 1960s and 1970s worked to make unpaid gendered labor visible and thus gain public recognition of its value, "this approach appears to transfer the temporal order of industrialization to caring labour, since it embeds caring labour in linear time and reshapes care as temporally distinct sets of activities.... This transfer cannot capture the 'highly fragmented' and overlapping character of some caring work... and may render some care invisible and meaningless against linear measurement and clock time pressures." This captures part of the friction between institutional time and personal care time at Sakura, while also pointing toward how care was interpreted by engineers in the development of care robots.

As we have seen, robotics engineers at the Robot Innovation Research Center (RIRC) took an approach of breaking care into what they assumed were its constituent parts, consisting of individual steps or tasks that could be organized into algorithms according to a temporal linearity: first this step, then that, then the following. Care robots were developed as materializations of these care algorithms, acting on aged bodies abstracted and rendered as computationally tractable packets of data—what one engineer at RIRC described as "a human, parameterized." Algorithms are not necessarily only computational or robotic: they also describe the increasingly rationalized processes, standard operating procedures, and routines through which *humans* perform care labor in ways that can appear as linearly organized, predictable, logistical steps to run through in sequence. Philosopher Mark Coeckelbergh contends that care robots are simply an instantiation of the modern organization of care labor in which care workers are increasingly treated as "cogs in a *care machine*" (2015, 276; original emphasis). In this way, the embodied algorithms both of care robots and of human care workers can be understood as part of a continuum of a techno-care assemblage. On first sight, the temporal regime of institutional care at Sakura certainly appeared algorithmic in the sense of following a set predetermined order—certain duties that had to be discharged at certain times. According to the logic at work at RIRC, if both robotic and human practices of care could be viewed as algorithms, then boundaries between human and nonhuman care could be overcome. Care could be rationalized by untangling its many threads and then run according to programmable logics. In theory, robots could eventually replace the human element.

Yet, viewing Sakura as a set of linear human/robot algorithms fails to capture the richness of the lived reality and meaning of care, care labor, and daily life. The processes of robot development such as those at RIRC involved a flattened perspective of life at the care home, achieved through abstracted and decontextualized analysis by engineers and software developers, giving the appearance of repetitive and therefore automatable routine. This flattening, however, erased the ethical values of care staff, including the nature of care practices that constructed relationships of mutual trust and ease over time by accumulating repeated social interactions between care workers and residents. It failed to take into account the fact that multiple temporalities overlapped, as did the individual duties carried out throughout the day—which could not always be separated out or made to be entirely linear. The "personal care" of personal care robots was different from the ethically valued personal care understood by care workers as requiring yoyū and slow communication.

Robotic Reconfigurations of Time and Space

How did using the robots impact the care home and its multiple temporalities? In reconfiguring care tasks, the introduction of care robots "in the wild" served to reconfigure space and time at Sakura in complex ways anticipated neither by their developers nor users. Although the robots in theory were designed for the discrete tasks of lifting, communication, and recreation, in actual use they intruded on overlapping areas of care and social life.

All three robots imposed new duties on both care workers and care recipients and were often described by staff as requiring the very same types of care that they were supposed to be providing to residents, almost as if they were residents themselves. Hug, the transfer robot, itself had to be "transferred" and wheeled around. Pepper, marketed as able to "watch over" residents, itself had to be watched over. Like many of the residents, it had to be assisted in moving around the floor, was in constant danger of falling or being accidentally pushed over and was assumed by care staff to be both precious and fragile. The therapeutic care that Paro provided was in fact generated from the caring response it was designed to evoke from users—it had to be cared for and worried about like a baby. Although intended to prevent wandering, cultivate a sense of psychological security, and increase social interaction, it was characterized by staff as leading to psychological wandering, unease, and even social withdrawal in the case of Ito. None of the devices could be left unattended without staff perceiving some kind of risk to residents.

Care workers interpreted the ethical value of the devices to be relational and revealed through their use, rather than simply as synonymous with the robots' physical safety or as a set of abstract moral rules programmed into them, as often understood by engineers. The way in which staff interpreted each device was based in large part on the manner in which it synchronized or failed to synchronize the three temporalities of the care home. These cases add further complexity to Yuji Sone's discussion of the "phenomenon of entrainment" (*hikikomi genshō*)—a concept he argues has shaped the development of care robots in Japan such as Paro. This idea refers to "the alignment of an organism's circadian rhythm with another's rhythm in a shared environment after a certain time spent together." As Sone explains,

The noun "hikikomi" relates to "dōka" (assimilation or adaptation) and "dōchō" (synchronisation, synchrony, sympathy, alignment, or conformity). What these concepts point toward is the synchronisation of the user and the robot, suggesting that robot design should take advantage of the fact that human users are capable of adapting themselves to objects in the everyday environment. (2017, 203)

Entrainment in the case of Pepper, Hug, and Paro, however, was not a straightforward process of users synchronizing themselves with each robot.

In requiring extra time to use, Hug slowed the process of lifting but reduced the amount of yoyū, or additional time/space available for care workers to provide personal care. In the case of Pepper, residents literally synchronized their actions with those of Pepper during rexercise. Although Pepper did not expressly cut down the opportunity for yoyū, it held the potential to do so by removing the need for verbal communication and spontaneous social interaction between care staff and residents during the crucial free time of recreation. Pepper also provided the clearest example of the ability of robots to mediate time and space from outside the care home. In enabling the spatiotemporal shifting of skills of care as well as of engineering and programming via its apps, Pepper opened up the possibility of deskilling care in the nursing home while facilitating the use of foreign care workers who may have fewer Japanese language skills.

Paro, however, provided an opposing case. Although intended to regulate mood and calm the user in order to smooth the institutional flow while maintaining "control" of the environment, it instead did the opposite. In transporting Ito, the resident who developed a strong attachment to Paro, into her own subjective experience of space-time, it disrupted the institutional routine, which perhaps helps explain why care workers seemed instinctively to hold back from taking it away from her. In catering to the user's self, and her own sense of time, Paro seemed a better fit with the ideology of personal care among staff. Yet, a strong element of ambivalence persisted due to the sense that this distancing from the institutional reality of Sakura exposed the impossibility of providing individual personal care to every resident within the institutional environment, while triggering a problematic retreat from the social life of the home.

In reconfiguring time, space, and practices of care, the use of the robots led to ethical negotiations among care workers, who often framed the use of the robots in moral gradations from ambivalence (Paro) through to rejection (Hug). Academic, media, and state discourses inside and outside Japan have long been concerned with the question of generalized public "acceptance" of robots—why Japanese people supposedly "accept" robots more readily than those in the West. Yet, interpreting the reactions of care workers and care recipients to the introduction of different types of robots is a far more complex and nuanced task that is highly dependent on context and use.

Robots and the Future of Care

Japan's growing care labor shortage, largely a result of the decisive national shift from familial to formal eldercare and accompanying explosion in demand for care workers combined with the low pay and socioeconomic status associated with such work, provided the rationale for a bloc of the Japanese government and industry to push the development and dissemination of care robots based on a logic of labor substitution—to provide a solution to the apparent contradiction of capitalism, or the idea that capitalism depends on the very social reproduction and care labor that it systematically undermines. Based on the data presented in this book, use of care robots such as the ones trialed at Sakura cannot currently, by itself, address the care labor deficit in Japan. There were three main reasons for this.

First, the robots did not decrease the amount of work for care staff but rather increased it, adding new tasks of setting up, configuring, moving, operating, mediating, storing, cleaning, maintaining, updating, managing, and overseeing them—in other words, taking care of them. Second, their use sometimes undermined the motivations of care staff who enjoyed the slow communication and tactile care vital to building meaningful relationships with residents. Use of all three robots could further reduce time and space for social contact between care workers and residents. Even leaving aside the likelihood that such a reduction in contact would be detrimental for residents, the impact on staff who are paid little more than the minimum wage and can leave for another job fairly easily could further worsen the already high industry-wide staff turnover rate and severe labor shortage. Finally, the robots were often simply too expensive for publicly funded institutions to buy, particularly at a time when changing legislation was speeding up the cycle of admissions to care homes and the intensity of care required by new residents. In the case of Sakura, by the end of the trial period, Mr. K decided that the expense of buying any of the robots was not worth the benefit. In this sense, care robots present a dual paradox. Technocrats in the government and METI intended them to act as cost-cutting, labor-saving devices to rein in the ballooning public expenditure on care. Yet, on the contrary, care robots cost additional money and require additional labor. This is a significant finding, given that governments globally, particularly in Northern Europe, are increasingly looking to robotic care for these benefits.

A corollary of the idea of robots filling the Japanese care labor gap was that they would prevent the need for (im)migrant care workers. But by reconfiguring the skills needed for key aspects of care that previously required Japanese language proficiency and the ability to build and maintain social relationships through tactile care and communication, the use of robots could help to facilitate the introduction of migrant care staff who may not possess these skills. By constructing an infrastructure of care devices that almost inevitably increases the amount of time taken for care tasks and creates new "invisible" tasks of robot operation and maintenance, it would seem to necessitate both a greater number of care workers, and lower-paid and lower-skilled care staff, to mitigate the huge expense that would be incurred by a national scaling-up of the implementation of such devices.

Skill and labor are reconfigured within the new assemblage of care that roboticization presupposes. The medium of the robot brings engineers and programmers to the care home, while extending the care home into robotics labs and programmers' offices—indeed, to individual programmers based anywhere in the world. The reconfiguration involved in robot care, with its expanded division of labor, points to an increased inequality and gendering of skill distribution—at one end of the scale are the highly abstract information technology skills of (predominantly male) robotics engineers and software developers, and at the other are (predominantly female) migrant workers who may not speak fluent Japanese or even have training in care but who could simply copy the bodily movements of the robot or do the manual labor of operating a robotic lifting device. This inequality of skill distribution in turn suggests the ongoing construction of an inherent power hierarchy with class, race, and gender implications as professional care labor could be deskilled and ultimately exposed to greater risk of precaritization.

Care robots in Japan have so far failed to live up to the expectations and imaginations of engineers, government officials, care home managers, care workers, and older users. One employee of an organization directly involved with the Robot Care Project pointed out to me despondently that the "robot society" imagined ten years earlier had not come to pass and posed the question, "So much government money has been poured in, but where did it all go?" He went on to argue that it had been largely wasted on too many small projects with unclear goals, with little to show for them. Despite the series of care robot policies, relatively few products have ended up in the marketplace. By October 2019, nineteen products had been developed and brought to market with the assistance of the Robot Care Project, after four years and at a cost of more than ¥12.5 billion ($125 million) to the government and at least the same amount again to the private sector, which was required to match or surpass the level of state investment. More broadly, the national government alone had spent well in excess of $300 million by the time of my research when we include various other development and popularization projects (Wright 2021), while the International Federation of Robotics reported that the entire global market for nursing care and disability aid robots stood at just $48 million in 2018 (International Federation of Robotics 2018)—less than one-tenth of the target of ¥50 billion ($500 million) by 2020 set out in the Japanese government's national robot strategy (Headquarters for Japan's Economic Revitalization 2015). Another report found that the size of the nursing care robot market in Japan in the same year was only ¥1.9 billion ($19 million) (Yano Research Institute 2018, cited in Tsujimura et al. 2020).

The extent to which these state efforts to date have actually succeeded in contributing to the development of either care robots or a market for them is difficult to assess. The companies involved may have developed them anyway without public support, while the products they did develop with the help of public funds may prove commercially unsuccessful. The apparently cohesive national strategy of robot care, like the unanimity of the Robot Care Project, was an illusion: the strategy was resisted by parts of the government and industry,[4] while the project through which it was implemented at RIRC represented a cobbling together of various personal research projects often unrelated to care. Many of the care robots themselves were repurposed from other uses: Hug was a repurposed robotic arm; Pepper was a commercial robot developed primarily to interact with customers in a retail environment, repurposed via a few apps for nursing care applications. To some extent, the very novelty or interpretive open-endedness that maintains a mystique around robots—a sense of the unknown—is also a reason for resistance to their adoption: because users in the care sector were not involved in their development, they seem strange, alienated from actual practices of daily life and care: expensive and irrelevant, unnecessary. Their uses or potential uses are not always immediately apparent; they were not designed for care.

Given the striking lack of meaningful collaboration between engineers and care workers and the minimal role played by either caregivers or care recipients in Japan's Robot Care Project, one of the main questions left hanging is what alternative forms of sociotechnical innovations might have been, and might yet be, developed if such collaborations were to happen. What if such innovations were developed without the explicit framing of trying to replace human caregivers or reduce public spending on care, without relying on abstract imaginings or negative stereotypes of older people and care work, and without starting from the assumption that the development of robots or other technologies alone holds the solution to the "problem" of care? What alternative futures are possible for older people, for their caregivers, and for care?

Care beyond Crisis: Coming to Terms with Aging

To provide affordable and sustainable care for people and to enable them to age comfortably and with dignity are not simply the aims of the Japanese welfare state, and the crisis of care is not simply a crisis of human eldercare. The chorus of scientists and activists warning about the accelerating climate emergency, as well as the rapid destruction of nonhuman habitats and the accompanying mass extinctions, irreversible loss of biodiversity, and increase in the frequency of pandemics, which has grown deafening in recent years, points to the need to allow the size of overconsuming human populations to fall as a significant (though far from sufficient) step toward preserving whatever degree of ecological sustainability is still salvageable.[5] If population sizes are to reduce, populations must age, and new ways of reconciling and rebalancing productivity and care—of people and planet—must be found. With a population estimated to shrink by more than half between 2017 and 2100 (Vollset et al. 2020), Japan can be seen as a prototype of how such aging could take place and in such a way as to reconceptualize substantial population reduction not as a national catastrophe but as a necessary pathway to human and nonhuman sustainable coexistence. The question then is how governments can maintain or improve quality of life by harmonizing—synchronizing—productive and reproductive labor, providing sufficiently rewarding jobs to care workers, and supporting formal and informal caring practices, processes, and institutions that can sustain the well-being of aging populations.

The Japanese experiment of creating a techno–welfare state for the twenty-first century through the hugely ambitious strategy of developing and implementing an infrastructure of robots aimed at carrying out tasks across almost every aspect of eldercare has so far failed to achieve its objectives. This was somewhat predictable—after all, the importance of involving end users actively and substantively at every stage of the conceptualization, development and implementation of new technologies, via practices of participatory or user-centered codesign, is clear (Bradwell et al. 2019; Fischer, Peine, and Östlund 2020).[6] Japan's pivot from familial care to a model of formal care supported by robots, if widely implemented, seems poised to lead not to a straightforward automation of care but rather to a deskilling, devaluing, and dehumanizing of care work and increasing distance between people giving and receiving care. Making safe, cheap, and effective robots to replace human care work remains a challenge that shows no signs of being accomplished anytime soon, despite the money being thrown at the attempt, particularly by Japanese and European governments. Robots are not yet the "solution" to the apparent contradiction of capitalism.

Finding alternative paths beyond such techno-fixes in Japan and the rest of the aging world will be challenging. It will involve gaining new, local understandings of care—not just of the physical practices but also of the situated meanings and ethics of care, by paying close attention to the genba (actual sites) where it is provided. The views of older people and their caregivers—the ultimate end users of care practices and arrangements—must be included in sociotechnical imaginaries about the future and in research and development processes. Above all, technologies must be seen not as stand-alone solutions universally applicable across every care setting but rather as components of contextually specific care assemblages. One problem with the algorithmic view of care and the idea that robots could substitute for care workers in the development practices at RIRC is that they seem to undermine the humanism of caring. Human caring relationships should be promoted instead of pared back and replaced or displaced by robotic or digital alternatives that serve to construct rather than prevent muen shakai—societies where meaningful personal ties have been severed. We need a far more critical approach to robotic systems that takes into account the whole sociotechnical assemblage of care they will inevitably reconstitute through their introduction.[7] We should call into question the economic logic of developing expensive technologies that can only be made affordable through economies of scale and a rationale of substitution or devaluation of human care labor. We should also urgently examine the ethical and environmental logics of using robots with short consumer lifespans and accompanying computational and data infrastructures made of materials and components that rely on highly polluting resource extraction and exploitative labor practices particularly in the Global South, which, if implemented on a mass scale, would generate even greater mountains of toxic electronic and plastic waste (Gabrys 2013; Meintjes 2020; Crawford 2021).

But Japan's attempts to implement robot care provide glimpses of alternative approaches. In the reaction of care workers to the imposition of robots, we find another model of personal care facilitated by yoyū—that spare capacity of time and space to care. Yoyū suggests a way to move toward more humanistic, caring rhythms that emphasize facilitating social interactions and developing social ties, making care "easier" for caregivers and care recipients by providing additional time to deliver personal care. This "solution" to the crisis of care is in some ways the most straightforward but also the hardest to accomplish under conditions of neoliberal capitalism: to acknowledge the skilled work involved in good care, make care a valued job, encourage more caregivers into the sector, resist the imposition of an industrialized temporality based on a logic of ever-greater productivity, and support and make financially and emotionally sustainable the formation of meaningful social relationships between those giving and receiving care, developing rather than discarding what Shannon Vallor (2015) calls the "moral skills" of care. It seems ironic that technologies such as robots are hailed as introducing "personalized care" as if this were an innovation, when it is human caregivers who individualize, personalize, and humanize care—even robot care—through their skilled social interactions with care recipients.

In the early 2020s, Japan is at a crossroads. On the one hand, the universal LTCI system seems to offer great opportunities. Despite the problems it has contributed to, including the care labor shortage, and despite the dangers of the overcommodification and marketization of care, or lack of sufficient or equitable provision in all cases, LTCI has made care work more visible and part of the paid economy. It has socialized the costs of eldercare and made care an accepted right. Japan has so far avoided both importing a large underclass of migrant female workers to do care labor and implementing expensive care robots on a large scale that could further undermine the value of care work—despite the best efforts of Abe, METI, and the robotics industry. The creation of the centralized apparatus of LTCI to carefully control and regulate the whole nationwide system of care assessment and provision, including in effect the salaries of care workers, offers the prospect of centrally raising these salaries, providing training to develop skills, and working with caregivers to identify and adopt effective existing technologies or develop new ones that can support rather than undermine their work. As the Women's Budget Group note, this kind of investment in the caring economy could improve the quality of care while providing an effective economic stimulus and reducing entrenched gender inequality (WBG 2020).

On the other hand, the push for care robotics in Japan, as in other parts of Asia and Europe, seems to be continuing, as does the neoliberal instinct to retrench welfare spending and "refamilize" care—pushing care responsibilities back onto individuals and close family members (Roberts and Costantini 2021). Growing numbers of older people who require care are being pushed to seek it privately outside of the LTCI system (Ogawa 2021). At the same time, the rapid increase in the number of foreign workers since 2000, together with the hurried deregulation of migration policy more recently and the entrance of many private recruitment agencies into this space, contributes to a sense that despite the delay caused by the closure of borders during the COVID-19 pandemic, it will be only a matter of time before far larger numbers of low-paid care staff from China and Southeast Asia are brought in. Meanwhile, care industry consolidation continues, supported by international capital.[8]

At the time of writing, Yoshihide Suga had resigned from his short tenure as prime minister, and it is unclear which direction his successor, Fumio Kishida, will pursue, particularly in the wake of the pandemic, which has thrown a spotlight in many countries on the future of long-term care while acting as a catalyst for the introduction of technocratic "solutions." Other countries are watching Japan closely: notably, in September 2021, the UK government proposed the introduction of a national insurance funding system for long-term care, partly inspired by the Japanese system, as part of the solution to its own care crisis (UK Government 2021). Across North America and Europe, however, valuing care is becoming ever harder within the current framework of financialized capitalism with its imperatives for productivity and profit extraction, and lack of accounting for the importance of skilled care labor. Governments in the European Union and the United Kingdom have invested record sums in innovative care technology projects including robots over the past decade while drastically cutting budgets for day-to-day social care services (Lipp 2019; Wright 2021). The "contradiction" of capitalism today seems to make the solution of better valuing, and paying, caregivers by definition impossible and suggests that the real innovation care requires is not primarily technological but rather social, political, and ideological—an innovation of capitalism itself.

Epilogue: Slippery Robots

It is difficult to come to grips with the temporalities of robots, whose public image of fast-paced innovation and relatively short time in the media spotlight contrast with the years or decades required for their development. Robots do not tend to age well, and while writing this book, I was warned that robots like Pepper would likely be obsolete by the time it was published and read, and that a new and better generation of care robots would quickly replace those discussed here. It is just a hundred years since the word *robot* was coined and only around twenty-five years since robots were first seriously proposed for use in eldercare. Although Japan may seem in stasis, with the government apparently incapable of bringing either robots or migrants decisively to bear on the continually growing shortage of care labor—a situation exacerbated by a pandemic that seemed to put everything indefinitely on hold—in fact, the country has undergone and is undergoing rapid change, particularly resulting from the revision of the labor laws in the late 1990s, the introduction of the LTCI system, and the sharp rise in the number of foreign workers since 2000. This book presents a snapshot in time and place, and no doubt other newer robots and high-tech devices will be developed and trialed and become part of the landscape of care in due course.

Yet, innovation in robotics happens slowly, in a complex interplay between the development of hardware and software, and robot time itself moves in nonlinear ways. Paro, first developed twenty years ago, still appears to new users as innovative and novel as it continues to travel the world. Pepper was launched in

Japan to great media fanfare in 2015, with the first several thousand units selling out within seconds of being released. In 2018, when the initial three-year leases started to come up for renewal, a survey found that 85 percent of Japanese businesses had decided not to renew (Tanaka 2018), and Peppers gradually started disappearing from the shops in big cities where they had once been a fairly common sight. But just at the time that it ceased to be a novelty in Japan, Pepper was being wheeled out in the United Kingdom as the epitome of futuristic high tech. In 2018, it was used to deliver a much-publicized prerecorded speech about the future role of robots to a British parliamentary committee. In 2019, it was touted by some English social care executives as the next big thing in care (ADASS 2019), and in 2020, the results of a research study that used Pepper for short stretches in nursing care in the United Kingdom and Japan received extensive and reverential news coverage (see, e.g., Khan 2020). Pepper was described in Euro-American media and even at international academic conferences as being commonly used in Japanese care homes, perhaps because of the regularity with which Pepper was (misleadingly) represented in media reports continuously amplifying the popular message of robots and Japan. At a major robot philosophy conference in 2020, Pepper was described by one speaker as "by far the most popular care robot in Japan." But in reality, by 2020, Pepper's main application in Japan seemed to be as a greeter at Hamazushi, a chain of inexpensive sushi restaurants. In the following year, SoftBank Robotics paused production of the robot and made significant cuts to its workforce. But robots like Pepper, even when they are no longer in production and fall into relative obsolescence, often continue to have a long tail—particularly via online media—playing a role in projecting and maintaining a techno-orientalist image of a futuristic Japan. This may in fact be their most successful role to date.

Perhaps the robots discussed in this book are simply at an early stage of development, and the next generation of care robots will overcome the challenges presented here. Perhaps advances in artificial intelligence will mean that they will require less human attention from care workers, maintain themselves, become safer. Perhaps their production will be scaled up, making them far more affordable and ubiquitous. Perhaps. But the phenomenal challenge of making robots that are safe, cheap, and effective can scarcely be overestimated: complexities multiply with the sophistication of the robot model, its capabilities, and the responsibilities it is given. The common yet deterministic way of thinking about robots as an ultimate end of capitalism presupposes proliferating sets of assumptions and speculation about possible futures, societal arrangements, and technological affordances.

One limitation of anthropological fieldwork is its fundamental slowness. Ethnography, like good care, requires yoyū. This can complicate the study of fast-moving technological developments and makes it difficult to speculate about the future. This is, however, also a strength, as it grounds analysis in the specificities of present conditions and forces us to look at what is in front of us—an antidote to the enchanting open-ended quality of robots. Robots are slippery in the sense that it is easy to find oneself slipping into the conditional, or even future, tense: if robots *were* cheaper, smarter, more reliable; robots *would* solve our problems; robots *will* save Japan. But it would seem unwise to pin our hopes on robots we expect to be developed in the future to solve challenges of eldercare already being faced today, particularly when alternative and more effective human-centered ways of providing care are far more readily available. Robotics, the quintessential modernist technology, is built on encouraging a propensity to imagine and believe in a better future robot, and a better future society, which does not currently, and may never, exist.

Notes

INTRODUCTION

1. This video promoted a free Pepper app called Haikai mimamori (literally, "Watching over [those who are] wandering") aimed at care homes and was posted on the website of its developer, NDSoftware Company, in 2016 (it has since been removed). It substantially overstates the actual capabilities of Pepper, which are discussed in detail in chapter 6.

2. These statistics are assembled from a number of sources: on demographic forecasts, see METI 2019; dependency ratio, see OECD 2017; on dementia rates, see Sado et al. 2018; numbers of care workers, see Statistics Bureau of Japan 2019 and Cabinet Office 2018; on labor figures, see Kawaguchi and Mori 2019 and www.tyojyu.or.jp/net/kenkou-tyoju/tyojyu-shakai-mondai/kaigojinzai-fusoku.html; and on social benefit spending, see Cabinet Office 2015 and Kato 2019.

3. Figures are from European Commission 2012a, 2012b, and U.S. Census Bureau 2017.

4. Independence is placed in scare quotes since this vision entails new dependencies on robotic devices and their accompanying technological and human infrastructures. Independence in the context of eldercare tends to suggest a sense of neoliberal responsibilization with the idea that people primarily have to help and care for themselves and their immediate family members, independent of social or political institutions (see Tronto 2013).

5. It is important to note that this economic "growth" largely represents activity that was already happening informally but is only now being made visible and counted as part of the formal economy.

6. The fact that the same title was also used by journalist Kishi Nobuhito for his 2011 book suggests the power of this narrative.

7. This is not to say that processes of roboticization or automation have ceased. On the contrary, the pandemic has acted as a catalyst for many information and communications technologies, including robots, while big tech companies have boomed (Ong 2020; Thomas 2020).

8. As Benjamin Lipp (2019) notes, Joseph Engelberger, one of the early pioneers of robotics, also imagined (albeit in somewhat vague terms) the potential future use of robots "aiding the handicapped and the elderly": "No, the robot will not be a practical nurse; but a robotized private abode will be so much more desirable than being in a . . . nursing home" (Engelberger 1989, 210–17).

9. Examples include SoftBank Robotics' Pepper (designed as a multipurpose humanoid), Sony's AIBO (designed as a substitute pet or toy), and the Korea Institute of Science and Technology's EngKey (an English-teaching robot from South Korea renamed Silbot for use in eldercare).

10. The concept of Society 5.0 was introduced in Japan's Fifth Science and Technology Basic Plan (2016–21) and continues to play a key role in the Sixth Basic Plan for the period 2021–26. It is defined as a "human-centered society that balances economic advancement with the resolution of social problems by a system that highly integrates cyberspace and physical space" (www8.cao.go.jp/cstp/english/society5_0/index.html).

11. Robertson uses a slightly differently formulation ("reactionary postmodernism, a nostalgic pastiche") to describe the depiction of the future Japan portrayed in the government document *Innovation 25* (2018, 57).

12. Use of the word *robot* in English actually predates its passage into more common usage and meaning from Čapek's play. The term, in this instance derived from a German word related to the Czech, was used in English texts in the nineteenth century to refer to the "central European system of serfdom, by which a tenant's rent was paid in forced labour or service" ("Robot, n.1." OED Online, June 2020).

13. This formulation is partially inspired by Beer, Fiske, and Rogers's definition (2014, 74) as well as by Dumouchel and Damiano 2017. Actuators are parts of a robot that move a mechanism (e.g., a robot arm); a manipulator is the robotic appendage (e.g., a hand or arm) that is used to physically manipulate materials.

14. One example was the use of Astro Boy (Tetsuwan atomu)—the robot boy who is perhaps the most popular anime character in Japan's history—as the brand ambassador to promote a major local robotics initiative in Kanagawa Prefecture called Robot Town Sagami. Part of the official website promoting the project was organized around the seven powers of Astro Boy, each of which was related to a real-life robotic device under development there (see http://sagamirobot.pref.kanagawa.jp/7powers.html). The use of pop culture references for robot branding is a common mode of state engagement—a way of mobilizing soft power to promote technocratic projects. Historian Ito Kenji has highlighted the events surrounding the celebration of Astro Boy's birthday in 2003 and the ways in which government and media used Astro Boy in "an attempt to redefine and re-create a culture of robotics in contemporary Japan by appropriating Japan's past cultural legacies. . . . Robotics engineers and robot manufacturers were able to use Astroboy [sic] to paint a very positive image of robots and future visions about relations between humans and robots" (2007, 1, 14).

15. Sheila Jasanoff and Sang-Hyun Kim define sociotechnical imaginaries as "collectively imagined forms of social life and social order reflected in the design and fulfillment of nation-specific scientific and/or technological projects" (2009, 120).

16. Rodney Brooks, the famous Australian roboticist who founded iRobot and Rethink Robotics, blogged a blistering critique of such practices. He writes of robots such as ASIMO, which he dismisses as "corporate marketing robots": "They are all fake! Fake in the sense that though they are presented as autonomous they are not. They are operated by a team of usually six people, off stage. And everything on stage has been placed with precision, down to the millimeter. . . . Those robots are not real" (Brooks 2017). Honda retired ASIMO in June 2018.

17. This is, in a sense, an inversion of Alan Turing's "imitation game" (popularly referred to as the Turing test), which involved asking whether a computer could imitate a human sufficiently well to deceive a human player (Turing 1950). The Wizard of Oz technique reverses the Turing test, in that the aim is, at least in part, to convince humans that robots are acting autonomously when they are in fact being controlled by humans.

1. CRISIS AND CARE ROBOTS

1. For example, Mori (1981), Allison (2006), Geraci (2006), and Robertson (2007), among others, have argued that social and particularly humanoid robots are more "accepted" in Japan due to animist beliefs rooted in Buddhism and/or Shinto, according to which all things (including robots) are imbued with spirit and can therefore be more easily understood and respected as valued entities with some kind of agency and personhood. This "techno-animist" explanation has begun to fall out of favor for a variety of reasons. Fabio Gygi (2018) and Yulia Frumer (2018) have critiqued the idea as confused, speculative, reductivist, and even nonsensical. For example, as Frumer notes, the techno-

animist argument is sometimes used to explain the popularity of humanoid robots in Japan, but "if people in Japan already felt that all objects had a spirit or sentience, then there would be no need to design machines that resembled the human form" (2018, 158). Moreover, there is little evidence that Japanese people are unique in how much they like robots (see, for example, Shibata et al. 2008 and Broadbent, Stafford, and MacDonald 2009).

2. The prevalent positive image of robots presented in Japanese media is actively encouraged by the huge advertising companies Dentsu and Hakuhodo. Dentsu in particular exerts powerful influence over the media (Gaulène 2016) and has a significant interest in Japanese robotics, having opened a "robot promotion center" in 2014. In recent years, this center has organized the sending into space of a small humanoid robot, Kirobo, and has promoted Matsukoroid, Ishiguro Hiroshi's android version of Japanese celebrity comedian Matsuko Deluxe, which starred in the popular TV show *Matsuko Matsuko*, was used in adverts for major companies and served as a public relations ambassador for the city of Sapporo (Kishimoto 2016). At the time of writing, the robot promotion center was working with Ishiguro and the National Museum of Emerging Science and Innovation (Miraikan) in Tokyo, while also promoting SoftBank Robotics' Pepper.

3. Despite the technocratic rhetoric of Abe's administration, it is worth recalling that in 2018, the minister responsible for cybersecurity policy was forced to admit that he had never actually used a computer (McCurry 2018).

4. Although such documents present aging per se as a social problem, the Japanese government continues to be dominated by older men. For example, the average age of former Prime Minister Suga's incoming cabinet in 2020 (composed of eighteen men and two women) was sixty, while Suga himself was seventy-one. This suggests a somewhat hypocritical ageism at the heart of the ruling Liberal Democratic Party administration.

5. It is important to note that prior to the Meiji period, different types of family and household had existed in Japan, varying by region, class, and social group (Ochiai 2005).

6. Cosima Wagner (2013, 172) notes that there were likely indirect and hidden employment impacts on female workers with little public voice. Moreover, Japanese roboticization of manufacturing may have contributed to job losses in other countries that could not compete on output, prompting anger particularly in the United States over Japanese companies "dumping" cheap products on the market.

7. Ueno's book cited above is titled *Ohitorisama no rōgo*—I translate this as *Alone in Old Age*, but *ohitorisama* is a polite way to say "one's own" or "group of one" and is the name given to a growing trend of individualized goods and services: there are now *ohitorisama* karaoke rooms (where you can sing karaoke by yourself), restaurants and bars (where everyone sits alone) and even, in rarer cases, weddings (where you can have a ceremony and honeymoon by yourself).

8. Indeed, even though the working age population dropped by more than 12 million people between 2000 and 2020, the overall size of the workforce *increased* by 814,000. This was partly due to the increase in the number of women working, as the female labor force participation rate grew steadily over this period, from 60 percent to 73 percent. Most of these women entering work, however, did not go into care jobs. Figures taken from the World Bank: https://data.worldbank.org/indicator/SP.POP.1564.TO?locations=JP, https://data.worldbank.org/indicator/SL.TLF.TOTL.IN?locations=JP, and https://data.worldbank.org/indicator/SL.TLF.ACTI.FE.ZS?end=2019&locations=JP&start=2000.

9. The government has also introduced policies aimed at supporting employees balancing work and caring responsibilities in order to reduce the number of people leaving the formal workforce to become unpaid carers, although uptake remains fairly low. Glenda Roberts and Hiroko Costantini (2021) argue that the paradigm of "balancing" work and care (*tomokea*) misses the fact that care and work are commonly understood as "competing devotions" and are therefore often seen as mutually incompatible, particularly by men. As

a result, people who feel overwhelmed by their dual responsibilities frequently quit their job to care full time even when their employer offers care leave and actively encourages their employees to take it.

10. Ito Peng and Joseph Wong caution against pigeonholing particular welfare states, including Japan, into a static common framework given their substantial differences and change over time, arguing that like some other East Asian nations, Japan has gradually moved toward a social insurance model that has "evolved into a set of programmes based on social solidarity, universality and with redistributive implications" (2010, 648). Nevertheless, a historically productivist or economic growth–focused orientation since the 1970s helps explain METI's ongoing technocratic interest in driving care policy.

11. www.jitco.or.jp/en/regulation/index.html

12. Across all of these migration channels, migrant care workers are currently allowed to work only in institutional care facilities—not in home care services or live-in care arrangements.

13. The name Inobe plays on the word *innovation* when transliterated into Japanese. For an excellent in-depth analysis of *Innovation 25*, see Robertson 2018.

14. The Japanese translation of ISO 13482's "personal care robot" is *seikatsushienrobotto* (literally, "lifestyle support robot"), although the term *kaigorobotto* (care robot) is also often used in government projects involving these devices, and the two terms seemed to be used interchangeably by engineers at AIST.

15. In 2015, the Japan Agency for Medical Research and Development (AMED) was created to bring all medical research projects across government under one roof. In the same year, funding oversight of the Robot Care Project was transferred from METI to AMED, despite the designation of care robots as nonmedical.

2. DEVELOPING ROBOTS AND DESIGNING ALGORITHMIC CARE

1. The organization of different research groups at AIST is complex and overlapping: RIRC, a research center subgroup of Intelligent Systems, overlapped with various research teams, with some team members also members of the Service Robotics Research Team, the RT Middleware Team (also called the Robot Software Platform Research Team), the Smart Mobility Research Team, the Dependable Systems Research Team, the Robot Software Research Laboratory, and the Human Informatics Research Team. Given this complexity, I avoid using confusing terminology in the description that follows and focus on the members of RIRC. Since the time of my fieldwork, the team has been renamed the Human Augmentation Research Center and relocated to AIST's newly constructed Kashiwa campus.

2. It is interesting to compare anthropologist Grant Otsuki's analysis of the role of silence at a wearable technology lab in Japan. He interprets it partly as a mark of deference of junior researchers toward senior professors and partly as their attempt to "reduce mental load" for senior lab members, while professors sometimes criticized the silence of juniors at research presentations as demonstrating disrespect or a lack of commitment to the lab (2015, 73–80). Anthropologists Anne Allison (2009) and Nana Gagné (2021) have shown how after-work social events provide important venues for (particularly male) office workers to communicate in a more direct manner with peers and superiors in a less constraining environment than the office. I attended only one such after-work event in Tsukuba, and this seemed to be a fairly rare occurrence for the team.

3. Respectively, the Japan Robot Association, Japan Assistive Products Association, Japan Quality Assurance Organization, Japan Standards Association, Japan Assistive Products Evaluation Center, the National Institute of Occupational Safety and Health, Japan, and the Association for Technical Aids.

4. For an in-depth analysis of this project and the shifting criteria against which its success or failure has been judged, see Garvey 2019.

5. It is also important to point out that other robotics engineers and research groups working on care technologies in Japan and elsewhere do use approaches that engage potential end users more actively and involve their closer participation in design and development, particularly as the benefits of such approaches have become clearer.

6. This event sparked something of a bemused backlash among Euro-American commentators, with headlines such as "Jerk Human Beats Up Boston Dynamics Robot" (https://mashable.com/2016/02/23/boston-dynamics-human-beats-up-atlas), as well as a satirical riposte in 2019 by U.S. production studio Corridor Digital: a digitally enhanced video in which a bullied robot from a fictional company called Bosstown Dynamics hits back against escalating acts of violence by its human tormentors (www.youtube.com/watch?v=dKjCWfuvYxQ).

7. It is important to note that robots for direct military applications are not developed in Japan due to its pacifist constitution. This was taken very seriously up to the period of my fieldwork. To give one example, a University of Tokyo robotics team was forbidden to compete at the United States' DARPA Robotics Challenge in 2013 because DARPA (the Defense Advanced Research Projects Agency) is a military organization; as a result, the team had to change its name to Team K and participate under the banner of the University of Hong Kong in order to enter the competition. Nevertheless, the situation becomes fuzzier in cases such as SoftBank's acquisition of the U.S.-based company Boston Dynamics, which had previously been funded by DARPA to develop several robots with potential military applications.

8. See the special issue on the ethics of robots in the Japanese-language journal *Society and Ethics* (*Shakai to rinri*, no. 28 [2013]). The importance of *kokoro* in Japanese robotics is taken up by Hirofumi Katsuno in his work on amateur robotics competitors (2010, 2011) and is also discussed by Yuji Sone (2017).

9. As Robertson (2018) points out, the Robot Revolution Realization Council was also dominated by industry figures, further indicating the influence of private industry over the direction of national robotics strategy.

10. The question of how to avoid Japanese robots becoming Galapagos technologies is addressed in Kishi Nobuhito's 2011 book titled *Robots Will Save Japan*.

11. "Algorithm, n.," OED Online, June 2020.

12. Rodney Brooks, who founded iRobot, the company that makes Roomba and also provided PackBot robots to the Japanese government to give emergency assistance in the wake of the 2011 Fukushima nuclear disaster, asserted, "Companion robots weren't any use in Fukushima. . . . And elderly people don't want companion robots. The elderly want control of their lives. They want dignity and they want independence. They don't want cute robots—it's about doing real tasks to make their lives easier" (Morris 2015).

3. PORTRAIT OF A CARE HOME

1. A day care center for older adults was located on the first floor, and there was a small day nursery for the children of staff members in the basement.

2. According to a national survey of 8,773 care workers conducted in 2016, women accounted for 73 percent, and men 25 percent (percentages do not add up to one hundred due to some nonresponses; Care Work Stabilization Center 2017).

3. Nationwide in the same year, the average hourly pay for care workers (before additional government wage subsidies) was ¥965, while the average minimum wage was ¥902 (Eggleston, Lee, and Iizuka 2021).

4. From 2015, prefectural governments began to introduce a limited subsidy, funded by the central government, for care institutions to buy certain models of care robot. The subsidy typically covered 50 percent of the cost up to a maximum of ¥100,000 ($1,000) per device. From 2018 (after the end of my fieldwork at Sakura), this ceiling was

increased to ¥300,000 ($3,000) per device. Some local governments also provided their own subsidy schemes.

5. Unlike the Japanese care workers who called each other by their surnames, Diego, like other foreign-born members of staff, was referred to and addressed by his first name despite having a Japanese surname.

6. Modern *enka* is a form of popular music in Japan usually consisting of sentimental ballads performed in an old-fashioned musical style.

7. Sakura was fairly representative of other nursing care homes in Japan in terms of the number and average age of care staff, their level of experience in the industry, and wages (MHLW 2017b).

8. This point stems from the work of Loretta Baldassar and colleagues on the kinning process between migrant care workers and older care receivers. See Baldassar, Ferrero, and Portis 2017.

9. This interpretation challenges some recent portrayals of Japanese care workers as "robotic," overly bureaucratic, and unable to express love, treat residents in familial ways, or engage effectively in affective labor. The latter depiction of Japanese care workers has been contrasted with, for example, supposedly more caring and familial Southeast Asian migrant care workers (Lopez 2012; Świtek 2014, 274; Lan 2018).

10. The verb (*sugosu*) often used in these contexts can be translated both as "to pass (time)" or by extension, "to live."

4. HUG: RECONFIGURING LIFTING

1. This chapter is derived in part from an article published in *Asian Anthropology* on January 4, 2018, © The Department of Anthropology, The Chinese University of Hong Kong, available online at www.tandfonline.com/doi/full/10.1080/1683478X.2017.1406576.

2. See, for example, a report by Japan's Association for Technical Aids (ATA 2015).

3. These used to be called no lifting policies. The first such piece of legislation was passed in the United Kingdom in 1992 in the form of the Manual Handling Operations Regulations, which obliges employers in all industries to avoid or reduce the need for employees to manually lift any load where there is a risk of injury. In the United States, there is no federal safe patient handling law, but such laws exist in around a dozen states. See www.cdc.gov/niosh/topics/safepatient/default.html.

4. This was exacerbated, particularly in the United States and Europe, during the COVID-19 pandemic, when nursing care homes became potentially fatal virus hotspots for staff as well as residents.

5. Both the Japan Industrial Safety and Health Association and the National Institute of Occupational Safety and Health, Japan have published guidance documents on preventing back pain among care workers. These efforts, however, have met with limited success, as most care homes in Japan continue the practice of manually lifting residents. An interviewee from the Japan Quality Assurance Organization told me that they were working with an organization in Japan to try to introduce such a policy, but that substantial inertial forces were at play, as complying with such a policy would require more time from care workers, increasing labor demands in an already fraught industry. They thought the government was therefore unlikely to introduce binding legal requirements.

6. This included lifting from bed to wheelchair, wheelchair to toilet, toilet back to wheelchair, wheelchair to bath, and so on, as well as lifting up residents who were slumping in their chairs.

7. This is a rough scale intended only to provide an indicative level of back pain.

8. It is important to add that staff members may have felt that back pain alone was not an "acceptable" reason to give for leaving their job. Mr. K's view, however, is supported by a national survey of care workers conducted in 2019, according to which 26 percent

of those who quit a previous care job said it was due to marriage, having a child, or doing childcare, 16 percent because of "problems with human relations in the workplace," and only 3 percent because of illness or old age (Care Work Stabilization Center 2020).

9. I had observed a handful of instances where care staff wheeled residents fairly quickly during busy times, although Mr. K's characterization here is clearly something of an exaggeration.

10. She also makes the important point that a caregiver cannot check the face of the person they are lifting for signs of discomfort while performing a manual transfer and hence argues that the use of lifting devices can enhance the quality of care and reduce any pain caused by manual lifting.

5. PARO: RECONFIGURING COMMUNICATION

1. Jennifer Robertson (2014, 2018) has pointed out that Paro was the first robot to be ceremonially granted its own *koseki* (household registry) certificate from the mayor of Nanto City in 2010. This also makes it the first robot to have received de facto national citizenship—a highly symbolic act, particularly given the extreme difficulty of even long-term immigrants obtaining citizenship in Japan.

2. Paro was classified as a Class II medical device in the United States in 2009 but currently holds a nonmedical CE (Conformité européene) certification in Europe. The National Advisory Committee for Assistive Devices has not yet included communication robots as a category of welfare equipment covered under the Long-Term Care Insurance system.

3. Shibata explained how this selection took place: After exhibiting a sixth-generation version of Paro at the Science Museum in London in 2002, he was approached by a researcher from Guinness World Records, who asked for evidence of therapeutic effects. Two weeks after sending some of his studies to the researcher, Shibata received confirmation that Paro would be certified as the world's most therapeutic robot.

4. The word *therapy* can be rendered in Japanese by its romanized loan word, *serapī*, by the rather vaguer word *iyashi* (healing), or with a more medical term such as *chiryō* (medical treatment).

5. It is important to note that given the short-term nature of most robot studies, it is often difficult to tell whether observed benefits of introducing a robot are simply caused by a novelty effect triggered by introducing a change into the care home environment. As Turkle has noted (2011, 105), it is likewise hard to distinguish older people's interest in robots from their interest in the researchers who come along to conduct the study.

6. This tweet was sent from Hirukawa's publicly available Twitter account (@hirohisah), accessed June 2016. All previous tweets from the account including this one have since been deleted, and the account itself appeared to have been transferred to a different user.

7. Doi Toshi, the inventor of Sony's popular AIBO dog-shaped robot, also described his creation as "completely useless" (www.economist.com/business/2000/12/21/dr-dois-useless-inventions).

8. Disasters have marked the introduction and accelerated adoption of iyashi, keichō (active listening), and various other mental health practices, sometimes in connection with "spiritual care" (McLaughlin 2013; Takahashi 2016). Robots such as Paro have also been deployed for use in the wake of disasters, particularly the 2011 Japanese earthquake and tsunami. The COVID-19 pandemic seems to have acted similarly as a catalyst for greater uptake of social robots—in this case, to provide the possibility for social interaction without the risk of passing on the virus.

9. Keichō courses, generally run by nonprofit organizations, have become an industry in themselves, with the price of certification, which is unregulated, ranging from around $200 for a course lasting a few hours to more than $1,500 for one spanning several days.

10. Keichō was not the only type of communication therapy or technique originally developed abroad that was being experimented with at Japanese care homes. For example, Humanitude®, a communication methodology and philosophy of care, developed specifically for eldercare in France in 1995 by Yves Gineste and Rosette Marescotti (Gineste and Pellissier 2007), has been popularized in Japan since the early 2010s by proponents such as Dr. Honda Miwako at the Tokyo Medical Center. This technique, focusing on eye contact, affective touch, and verbal communication to convey dignity and respect for the care recipient, was being trialed at a number of care homes, including one in Ibaraki Prefecture that was (separately) collaborating with AIST on the Robot Care Project.

11. Communication therapy techniques and technologies mentioned in this chapter—including keichō and iyashi as well as Paro and SimSensei—tend to be trialed or employed in a variety of mental health use cases that are believed to be improved through communication, including psychological trauma, dementia, PTSD, loneliness, depression, chronic pain, autism, and others. To some extent, this reflects a heuristic approach to finding new applications for emerging therapies, yet it also perhaps suggests a conflation of very dissimilar mental health conditions by engineering or computer science researchers.

12. It is important to recall that this researcher's home institute was partly funded by the U.S. military. According to the National Comorbidity Survey Replication, conducted between 2001 and 2003, 3.6 percent of all adults in the United States (or approximately 5.2 million people) had suffered from PTSD at some point in the previous year; the proportion within the veteran population was far higher. See https://www.nimh.nih.gov/health/statistics/post-traumatic-stress-disorder-ptsd.shtml.

13. Leeson reaches a similar conclusion in her analysis of Telenoid, a Japanese teleoperated robot with abstracted humanoid features that literally "dehumanizes" the appearance of the care worker who is remotely operating it, enabling an affective and confessional mode of communicative engagement with residents that had previously been impossible. She describes Telenoid as "a distancing tool that simultaneously allowed staff to engage in new forms of care and intimacy" (2017, 233).

14. The suffix—*kun* is usually used for boys. Paro did not appear to be assigned a consistent gender by residents or staff.

15. Incidents of users crying have been reported during other trials of Paro and similar social robots in eldercare by Cathrine Hasse (2013) and Christina Leeson (2017) in Denmark, and Sherry Turkle and other researchers from MIT (2006) in the United States. A professor at the University of Texas at Tyler reported one such reaction of a patient with dementia who used Paro: "It made her cry at first, which was a positive thing in her case because there was no evidence of emotion prior to that. . . . As she cried, she clung to the seal like she wasn't going to let go of it, so we just let her hold it. It evidently had tapped into some memory of hers from years ago, like of a long lost pet or something" (University of Texas at Tyler Magazine 2016). The response of care staff to such affective reactions from users differs: whereas the American author of the paper interpreted it in a positive light, Japanese care workers expressed ambivalence, which may have been related to more guarded cultural attitudes toward public displays of emotion. A similar case described by Hasse (2013) led to a reorganization of work routines due to the need for more careful observation from care staff of residents' interactions with Paro.

16. The word *ko* in Japanese can be used to mean "child" or jokingly, "pet." It is unclear which Ito meant here.

17. On the ongoing debate over the morality of deception in dementia care, see MacFarquhar 2018.

6. PEPPER: RECONFIGURING RECREATION

This chapter is derived in part from an article published in *Critical Asian Studies* on May 10, 2019, © The Bulletin of Concerned Asian Scholars Inc, available online at www.tandfonline.com/10.1080/14672715.2019.1612765.

1. Since the term *care robots*, potentially comprising a large range of heterogeneous devices, was not clearly defined in these polls and since the majority of those questioned would likely not have had direct experience with or even seen such a robot in real life, the results have been far from convincing.

2. Alphabet made a major move into robotics in 2013 under the guidance of Andy Rubin (the former head of Android), acquiring nine of the world's leading robot companies, before abandoning the plan to build a robot division when Rubin left the company. Having bought Boston Dynamics from Alphabet for a reported $100 million in 2017, SoftBank sold a controlling stake in the company to Hyundai in 2020 for $880 million—the steep rise in value an indication of the explosion in hype around robotics and artificial intelligence during this period.

3. As described, for example, by SoftBank Robotics' chief business officer, Yoshida Kenichi (www.japansociety.org/event/meet-pepper-the-worlds-first-humanoid-robot-that-reads-human-emotions).

4. https://vimeo.com/175578631

5. www.youtube.com/watch?v=OX2-tpTkLA0

6. In fact, in a 2015 TV advert for Boss Coffee that features Pepper, the robot denies being able to do eldercare at all (see www.youtube.com/watch?v=dcX9QHVlArk).

7. Two outwardly identical versions of Pepper were marketed and available for lease at the time of my fieldwork: "Pepper," which had relatively basic functions, and "Pepper for Biz," which was aimed at businesses and had more capabilities and a larger range of available apps. The unit provided for the trial at Sakura was the latter.

8. A case in point was the Wandering Monitoring (Haikai mimamori) app, developed by NDSoftware Company. A promotional video for the free app, briefly described at the start of this book, indicated that Pepper could identify older people wandering care home corridors at night and keep them talking while sending an email to alert a care worker. As staff at SoftBank's own "atelier" in Tokyo informed me, this app was ineffective for several reasons. Pepper could overheat if it moved around too much, so for the app to work, it would have to be positioned in a stationary vantage point. The lighting would have to be bright enough for Pepper to see clearly the face of the older man—therefore, the corridor lights would have to be left on, making it harder for residents to sleep. The man would have to stop exactly three feet away from the front of Pepper, look directly at its camera to show his face clearly, and then wait for Pepper to verify his face and allow Pepper to keep him talking while staff received an email telling them what was happening. This assumed that Pepper did not run out of battery or stop working for some reason, as it frequently did, or fail to correctly recognize the user's face, that the older man did not accidentally bump into or trip over Pepper, and that the staff were constantly monitoring their email (and thus, presumably, out of visual range of the corridor). It also presupposed that the man and his relatives would agree to images of his face being uploaded to the cloud for Pepper to access. Staff at the atelier told me that actually trying to use Pepper in this capacity would create many more risks than it mitigated. The app seemed more a marketing ploy than a serious tool for care.

9. Ritsuko's Rexercise was developed by Fubright Communications using a recreational exercise routine developed by Yamazaki Ritsuko, professor of nursing and managing director of an exercise research center.

10. A growing number of studies are similarly finding that using robots tends to create more work for caregivers. See, e.g., Demange, Pino, Rigaud, and Cantegreil-Kallen 2019; Huisman and Kort, 2019; Melkas, Hennala, Pekkarinen, and Kyrki 2020; Nickelsen 2019; Read, Woolsey, McGibbon, and O'Connell 2020 (all cited in Persson, David Redmalm & Clara Iversen 2021); Leeson 2017; Vogt and König 2021.

11. This migration channel was also hit by a scandal in March 2019, when seven hundred foreign students enrolled at the Tokyo University of Social Welfare were found to have stopped attending classes and disappeared. The government increased its scrutiny of foreign students following this incident, and Mr. K told me that the Ministry of Justice subsequently rejected the entry into Japan of several students he had been hoping to hire.

12. In the United Kingdom in 2019, a director of adult services at one of England's local councils that was also trialing Pepper in eldercare said, "People often say we're not trying to replace people with technology but what if we are? In Southend, we're trying to introduce that into the debate with our work with the Pepper robot. If Pepper had arms [sic] and legs then, yeah, we probably would be using it to replace humans in some way" (ADASS 2019).

13. Providing highly insightful parallel cases, sociologist Judy Wajcman (2015, 2016) has noted the complex ways in which the introduction of domestic technologies has "shifted" rather than reduced the time spent on housework, as well as the similarly hidden labor tasks involved in maintaining and upgrading digital infrastructure in the home.

14. Núria Vallès-Peris and Miquel Domènech define care fragmentation as "the division of caring processes and relations into different sorts of care, which are then provided by people from different professions, by different organizational services, or by different devices" (2020, 172). They note that this kind of fragmentation is becoming more frequent as implementations of care technologies accelerate across higher income countries. A similar point is made by Nelly Oudshoorn, who found that the introduction of telecare technologies for heart patients entailed a similar rise in the number of actors involved in health care and, as with care robots, increased rather than reduced the need for human labor (2011, 190).

15. The report puts a positive gloss on these findings, however, with the authors arguing that by encouraging more flexible work without taking jobs, robots can reduce the burden on care workers, although no qualitative data are provided to back up this hypothesis.

7. BEYOND CARE ROBOTS

1. The film was a hit on its release, and its title was even adopted as the name of a magazine, *Les temps modernes*, founded by philosophers including Jean-Paul Sartre, Simone de Beauvoir, Maurice Merleau-Ponty, and Raymond Aron, which was first published in France in 1945 and became a focal point of postwar French intellectual life (Appignanesi 2005, 82). When I asked Maeda why he showed *Modern Times* rather than old Japanese movies, he told me that many residents did not like Japanese films from that era, which tended to be "dark" and reminded them of the war or hard times during the postwar period. He said that many residents liked Charlie Chaplin. The silent movie format worked well, as many residents were hard of hearing but could read the intertitles and appreciated Chaplin's physical comedy (another favorite was Mr. Bean).

2. For a detailed discussion of the political context and themes of *R.U.R.*, see Richardson 2015, 24–38.

3. In a meeting in October 2019, the RIRC team leader Matsumoto Yoshio, who had been closely involved for several years in research about the care robot market in Japan, told me that he thought even a figure of 10 percent was likely an overestimate. Adoption in home care seems even lower: a 2021 study conducted in Japan found that out of a sample of 444 professional home care workers less than 1 percent reported having used a robot (Ide et. al 2021).

4. It is notable that some of Japan's largest technology companies chose not to participate in the Robot Care Project, with some interviewees from these companies expressing wariness or outright criticism of such government-run innovation projects (see also Lechevalier, Ikeda, and Nishimura 2010). As the head of the service robot division of a large automobile and technology company told me in 2019, "If we collaborate with government, there are many problems—it's a hassle! There are many limitations to development if we collaborate with the government, so we don't."

5. This is not to suggest that human population size alone is the determining factor in environmental catastrophe, which many see also as the result of extractive capitalism and its imperatives and incentives to exploit the environment and consume resources in unsustainable ways. Nevertheless, as Haraway puts it, "a 9 billion increase of human beings over 150 years, to a level of 11 billion by 2100 if we are lucky, is not just a number" (2016, 7). Even this expected peak in human population size (as predicted in the United Nations' World Population Prospects 2019) and subsequent decline is far from inevitable: many governments, including Japan's, are actively pursuing pronatalist policies in attempts to raise their populations' total fertility rate over the coming decades.

6. The importance of actively involving users in research and development processes has of course been known for decades. I found many echoes of the approach at RIRC documented in the work of anthropologist Diana Forsythe, who studied U.S. artificial intelligence labs developing expert systems in the 1980s and early 1990s: "Designed, built, and evaluated according to procedures that 'delete the social' and mute the voices of users, most of these systems remain 'on the shelf,' a fact which is hardly surprising" (2001, 15).

7. One promising approach is Aimee van Wynsberghe's care centered value-sensitive design methodology for service robot development, which is grounded in care ethics and focuses on the relational nature of care activities (van Wynsberghe 2016).

8. The 2010s saw the rise of new financial vehicles called real estate investment trusts (REITs), usually backed by international institutional investors, which have been used to build and run care homes in various countries including the United States and United Kingdom. REITs entered the Japanese market in 2013, following the recommendation of a government-backed panel which argued that they could help fund the construction of new eldercare facilities. At the time of writing, there were at least four REITs in operation in Japan, and these have been involved in a wave of acquisitions of care facilities. The eldercare market in Japan is seen by international investors as a large, stable, and growing source of profit, particularly in light of the trillions of yen in public investment being pumped into it via LTCI (O'Dea 2017; Savills 2019. I would like to thank Amy Horton for these references). REITs have acquired a reputation for aggressively maximizing and extracting profit by producing large, standardized, hotel-like care homes in order to achieve economies of scale, raising charges for residents but lowering wages for staff, and minimizing costs through sparing use of care supplies (Horton 2020).

References

Acceptable Intelligence with Responsibility (AIR). 2018. "Perspectives on Artificial Intelligence/Robotics and Work/Employment." http://sig-air.org/wp/wp-content/uploads/2018/07/PerspectivesOnAI_2018.pdf.
Allison, Anne. 2006. *Millennial Monsters*. Berkeley: University of California Press.
Allison, Anne. 2009. *Nightwork*. Chicago: University of Chicago Press.
Allison, Anne. 2013. *Precarious Japan*. Durham, NC: Duke University Press.
Ambo, Phie. 2007. *Mechanical Love*. Denmark.
Appignanesi, Lisa. 2005. *Simone de Beauvoir*. London: Haus.
Asis, Edward, and Rogie Royce Carandang. 2020. "The Plight of Migrant Care Workers in Japan: A Qualitative Study of their Stressors on Caregiving." *Journal of Migration and Health* 1–2 (2020), 100001. https://doi.org/10.1016/j.jmh.2020.100001.
Association of Directors of Adult Social Services (ADASS). 2019. "Connected Social Care: Adult Social Care Leaders Explore the Pros, Cons and Challenges of Tech-Enabled Care." New Dialogues, Spring 2019. www.adass.org.uk/media/6806/adass-tunstall-round-table.pdf.
Association for Technical Aids (ATA). 2015. "Kaigorobotto no kaihatsujōkyō to katsuyōsuishin ni tsuite" [On the state of the development and promotion of the use of care robots]. www.mhlw.go.jp/file/05-Shingikai-12201000-Shakaiengokyokushougaihokenfukushibu-Kikakuka/0000076872.pdf.
Baldassar, Loretta, Laura Ferrero, and Lucia Portis. 2017. "'More like a Daughter than an Employee': The Kinning Process between Migrant Care Workers, Elderly Care Receivers and Their Extended Families." *Identities* 24 (5): 524–41. https://doi.org/10.1080/1070289X.2017.1345544.
Beer, Jenay M., Arthur D. Fiske, and Wendy A. Rogers. 2014. "Toward a Framework for Levels of Robot Autonomy in Human-Robot Interaction." *Journal of Human-Robot Interaction* 3 (2): 74–99.
Berthin, Michael. 2014. "'Touch Future x Robot': Examining Production, Consumption, and Disability at a Social Robot Research Laboratory and a Centre for Independent Living in Japan." DPhil diss., London School of Economics.
Bethel, Diana Lynn. 1992. "Life on Obasuteyama, or, Inside a Japanese Institute for the Elderly." In *Japanese Social Organisation*, edited by Takie Sugiyama Lebra, 109–34. Honolulu: University of Hawai'i Press.
Bradwell, Hannah L., Katie J. Edwards, Rhona Winnington, Serge Thill, and Ray B. Jones. 2019. "Companion Robots for Older People: Importance of User-centred Design Demonstrated through Observations and Focus Groups Comparing Preferences of Older People and Roboticists in South West England." *BMJ Open* 9:e032468. http://doi.org/10.1136/bmjopen-2019-032468.
Bremner, Brian. 2015. "Japan Unleashes a Robot Revolution: Its Domination of the Industry Is Challenged by Korea and China." Bloomberg Businessweek, May 28, 2015. www.bloomberg.com/news/articles/2015-05-28/japan-unleashes-a-robot-revolution.
Broadbent, E., R. Stafford, and B. MacDonald. 2009. "Acceptance of Healthcare Robots for the Older Population: Review and Future Directions." *International Journal of Social Robotics* 1:319–30.

Brooks, Rodney Allen. 2017. "Domo Arigato Mr. Roboto." August 28, 2017. https://rodneybrooks.com/forai-domo-arigato-mr-roboto/.
Brucksch, Suzanne, and Franziska Schultz. 2018. *Ageing in Japan. Domestic Healthcare Technologies*. Tokyo: Leiden Asia Centre (LAC) and German Institute for Japanese Studies (DIJ). www.dijtokyo.org/wp-content/uploads/2018/09/Final_Report-30.08.2018.pdf.
Brynjolfsson, Erik. 2014. *The Second Machine Age: Work, Progress, and Prosperity in a Time of Brilliant Technologies*. Edited by Andrew McAfee: New York: Norton.
Buch, Elana D. 2015. "Anthropology of Aging and Care." *Annual Review of Anthropology* 44:277–93.
Cabinet Office, Government of Japan. 2015. "Annual White Paper on the Aging Society." www8.cao.go.jp/kourei/english/annualreport/index-wh.html.
Cabinet Office, Government of Japan. 2018. "2040 nen wo misueta shakaihoshō no shōrai mitōshi" [The future outlook for social security focusing on the year 2040]. www5.cao.go.jp/keizai-shimon/kaigi/minutes/2018/0521/agenda.html.
Cabinet Office, Government of Japan. 2019. "Cross-ministerial Strategic Innovation Promotion Program (SIP) Research and Development Plan for Big-Data and AI-Enabled Cyberspace Technologies." www8.cao.go.jp/cstp/english/01_cyber_rdplan.pdf.
Campbell, John Creighton. 1992. *How Policies Change: The Japanese Government and the Aging Society*. Princeton, NJ: Princeton University Press.
Care Work Stabilization Center. 2017. "Heisei 28 nendo kaigorōdōjittaichōsa no kekka" [Results of the 2016 care work conditions survey]. www.kaigo-center.or.jp/report/h28_chousa_01.html.
Care Work Stabilization Center. 2020. "Reiwa gannendo kaigorōdōjittaichōsa no kekka" [Results of the 2019 care work conditions survey]. www.kaigo-center.or.jp/report/2020r02_chousa_01.html.
Clifford, James. 1996. "Anthropology and/as Travel." *Etnofoor* 9 (2): 5–15.
Coeckelbergh, Mark. 2015. "Artificial Agents, Good Care, and Modernity." *Theoretical Medicine and Bioethics* 36:265–77. https://doi.org/10.1007/s11017-015-9331-y.
Crawford, Kate. 2021. *The Atlas of AI: Power, Politics, and the Planetary Costs of Artificial Intelligence*. New Haven, CT: Yale University Press.
Curtis, Adam. 2016. *HyperNormalisation*. BBC.
Danely, Jason. 2016. "Learning Compassion: Everyday Ethics among Japanese Carers." *Inochi no mirai* [The future of life] 1:170–92.
Deleuze, Gilles. 1989. *Cinema 2: The Time Image*. Translated by Hugh Tomlinson and Robert Galeta. Minneapolis: University of Minnesota Press.
Demange, M., M. Pino, H. Kerhervé, A.S. Rigaud, and I. Cantegreil-Kallen. 2019. "Management of Acute Pain in Dementia: A Feasibility Study of a Robot-assisted Intervention." *Journal of Pain Research* 12: 1833–46. https://doi.org/10.2147/JPR.S179640.
DeWit, Andrew. 2015. "Komatsu, Smart Construction, Creative Destruction, and Japan's Robot Revolution." *Asia-Pacific Journal* 13 (5). https://apjjf.org/2015/13/5/Andrew-DeWit/4266.html.
Donath, Judith. 2020. "Ethical Issues in Our Relationship with Artificial Entities." In *The Oxford Handbook of Ethics of AI*, edited by Markus D. Dubber, Frank Pasquale, and Sunit Das, 53–73. Oxford: Oxford University Press.
Dumouchel, Paul. 2016. "Feeling Robots and Feelings about Robots." Paper presented at the Anthropology of Japan in Japan conference, University of Tsukuba, November 26–27.
Dumouchel, Paul, and Luisa Damiano. 2017. *Living with Robots*. Translated by Malcolm DeBevoise. Cambridge, MA: Harvard University Press.

Eggleston, Karen, Yong Suk Lee, and Toshiaki Iizuka. 2021. *Robots and Labor in the Service Sector: Evidence from Nursing Homes*. NBER Working Paper Series. Cambridge, MA: National Bureau of Economic Research. www.nber.org/papers/w28322.
Emont, Jon. 2017. "Japan Prefers Robot Bears to Foreign Nurses." *Foreign Policy*, March 1, 2017.
Engelberger, Joseph. 1989. *Robotics in Service*. Cambridge, MA: MIT Press.
European Commission. 2012a. *Ageing Report: Europe Needs to Prepare for Growing Older*. Brussels: European Commission. http://ec.europa.eu/economy_finance/articles/structural_reforms/2012-05-15_ageing_report_en.htm.
European Commission. 2012b. *Long-Term Care for the Elderly: Provisions and Providers in 33 European Countries*. Brussels: European Commission. https://op.europa.eu/en/publication-detail/-/publication/6f79fa54-1199-45e0-bbf3-2619c21b299a.
Fischer, Björn, Alexander Peine, and Britt Östlund. 2020. "The Importance of User Involvement: A Systematic Review of Involving Older Users in Technology Design." *The Gerontologist* 60 (7): e513–e523. http://doi.org/10.1093/geront/gnz163.
Folbre, Nancy. 2001. *The Invisible Heart: Economics and Family Values*. New York: New Press.
Ford, Martin. 2015. *Rise of the Robots: Technology and the Threat of a Jobless Future*. New York: Basic Books.
Forsythe, Diana E. 2001. *Studying Those Who Study Us: An Anthropologist in the World of Artificial Intelligence*. Stanford: Stanford University Press.
Foster, Malcolm. 2018. "Robots Making Inroads in Japan's Elder Care Facilities, but Costs Still High." *Japan Times*, March 30, 2018. www.japantimes.co.jp/news/2018/03/30/national/robots-making-inroads-japans-elder-care-facilities-costs-still-high/.
Fraser, Nancy. 2016. "Contradictions of Capital and Care." *New Left Review* 100. https://newleftreview.org/issues/ii100/articles/nancy-fraser-contradictions-of-capital-and-care.
Frey, Carl Benedikt, and Michael A. Osborne. 2013. "The Future of Employment: How Susceptible Are Jobs to Computerisation?" Oxford Martin School Programme on the Impacts of Future Technology, University of Oxford.
Frumer, Yulia. 2018. "Cognition and Emotions in Japanese Humanoid Robotics." *History and Technology* 34 (2): 157–83. http://doi.org/10.1080/07341512.2018.1544344.
Gabrys, Jennifer. 2013. *Digital Rubbish: A Natural History of Electronics*. Ann Arbor: University of Michigan Press.
Gagné, Nana Okura. 2021. *Reworking Japan: Changing Men at Work and Play under Neoliberalism*. Ithaca, NY: Cornell University Press.
Garvey, Colin. 2019. "Artificial Intelligence and Japan's Fifth Generation: The Information Society, Neoliberalism, and Alternative Modernities." *Pacific Historical Review* 88 (4): 619–58. http://doi.org/10.1525/phr.2019.88.4.619.
Gaulène, Mathieu. 2016. "Does the Advertising Giant Dentsu Pull the Strings of the Japanese Media?" *Asia-Pacific Journal: Japan Focus* 14 (11). https://apjjf.org/2016/11/Gaulene.html.
Genda, Yuji. 2017. *Hitodebusoku na no ni naze chingin ga agaranai no ka?* [Why do wages not increase despite the labor shortage?]. Tokyo: Keio University Press.
Geraci, Robert M. 2006. "Spiritual Robots: Religion and Our Scientific View of the Natural World." *Theology and Science* 4 (3): 229–46.
Gibbs, Samuel. 2017. "The Future of Funerals? Robot Priest Launched to Undercut Human-led Rites." *The Guardian*, August 23, 2017. www.theguardian.com/technology/2017/aug/23/robot-funerals-priest-launched-softbank-humanoid-robot-pepper-live-streaming.

Gineste, Yves, and Jérôme Pellissier. 2007. *Humanitude: Comprendre la vieillesse, prendre soin des hommes vieux* [Humanitude: Understanding old age, caring for old men]. Paris: Armand Colin.

Gordon, Andrew. 2017. "New and Enduring Dual Structures of Employment in Japan: The Rise of Non-regular Labor, 1980s–2010s." *Social Science Japan Journal* 20 (1): 9–36. http://doi.org/10.1093/ssjj/jyw042.

Government of Japan. 2016. *The 5th Science and Technology Basic Plan*. Tokyo: Government of Japan.

Gygi, Fabio. 2018. "Robot Companions: The Animation of Technology and the Technology of Animation in Japan." In *Rethinking Relations and Animism: Personhood and Materiality*, edited by Miguel Astor-Aguilera and Graham Harvey, 94–111. London: Routledge.

Haraway, Donna J., 2016. *Staying with the Trouble: Making Kin in the Chthulucene*. Durham, NC: Duke University Press.

Hashimoto, Akiko, and John W. Traphagan. 2008. *Imagined Families, Lived Families: Culture and Kinship in Contemporary Japan*. Albany: State University of New York Press.

Hasse, Cathrine. 2013. "Artefacts That Talk: Mediating Technologies as Multistable Signs and Tools." *Subjectivity* 6 (1): 79–100.

Headquarters for Japan's Economic Revitalization, Government of Japan. 2015. "New Robot Strategy: Japan's Robot Strategy." Ministry of Economy, Trade, and Industry. www.meti.go.jp/english/press/2015/0123_01.html.

Holliday, Ian. 2000. "Productivist Welfare Capitalism: Social Policy in East Asia." *Political Studies* 48 (4): 706–23.

Horton, Amy. 2020. "Liquid Home? Financialisation of the Built Environment in the UK's 'Hotel-style' Care Homes." *Transactions of the Institute of British Geographers* 46 (1): 179–92. https://doi.org/10.1111/tran.12410.

Howell, Signe. 2003. "Kinning: The Creation of Life Trajectories in Transnational Adoptive Families." *Journal of the Royal Anthropological Institute* 9 (3): 465–84.

Huisman Chantal, and Helianthe Kort. 2019. "Two-Year Use of Care Robot Zora in Dutch Nursing Homes: An Evaluation Study." *Healthcare (Basel)* 7 (1): 31. https://doi.org/10.3390/healthcare7010031.

Hurst, Daniel. 2018. "Japan Lays Groundwork for Boom in Robot Carers." *The Guardian*, February 6, 2018. www.theguardian.com/world/2018/feb/06/japan-robots-will-care-for-80-of-elderly-by-2020.

Ide, Hiroo, Naonori Kodate, Sayuri Suwa, Mayuko Tsujimura, et al. 2021. "The Ageing 'Care Crisis' in Japan: Is there a Role for Robotics-based Solutions?" *International Journal of Care and Caring* 5 (1): 165–71. https://doi.org/10.1332/239788220X16020939719606.

International Federation of Robotics. 2018. *World Robotics Service Robots Report*. Frankfurt: International Federation of Robotics.

International Organization for Standardization (ISO). 2014. "13482: Robots and Robotic Devices—Safety Requirements for Personal Care Robots." Geneva: International Organization for Standardization.

Ishiguro, Nobu. 2018. "Care Robots in Japanese Elderly Care: Cultural Values in Focus." In *The Routledge Handbook of Social Care Work around the World*, edited by Karen Christensen and Doria Pilling, 256–70. London: Routledge.

Ishihara, Kohji. 2014. "Roboethics and the Synthetic Approach—A Perspective on Roboethics from Japanese Robotics Research." In *Robotics in Germany and Japan: Philosophical and Technical Perspectives*, edited by Michael Funk and Bernhard Irrgang, 45–58. Berlin: Peter Lang.

Ito, Kazuya. 2021. "Japan Sees 6-fold Rise in Number of Foreign Workers on New Skills Visa." *Asahi Shimbun*, May 26, 2021. www.asahi.com/ajw/articles/14358579.

Ito, Kenji. 2007. "Astroboy's Birthday: Robotics and Culture in Contemporary Japanese Society." Paper presented at the East Asian Science, Technology and Society conference, Taipei, August 7.

Iwabuchi, Koichi. 2002. *Recentering Globalization: Popular Culture and Japanese Transnationalism*. Durham, NC: Duke University Press.

Japan Robotic Association/Japan Machinery Federation (JARA/JMF). 2001. *Summary Report on Technology Strategy for Creating a 'Robot Society' in the 21st Century*. Tokyo: Japan Robotic Association/Japan Machinery Federation.

Jasanoff, Sheila, and Sang-Hyun Kim. 2009. "Containing the Atom: Sociotechnical Imaginaries and Nuclear Power in the United States and South Korea." *Minerva* 47 (119). https://doi.org/10.1007/s11024-009-9124-4.

Jozuka, Emiko. 2018. "Beyond Dimensions: The Man Who Married a Hologram." CNN, December 29, 2018. https://edition.cnn.com/2018/12/28/health/rise-of-digisexuals-intl/index.html.

Kageki Noriko. 2012. "Mujinbakugekiki no riyō wa beikokukenpō ni ihan suru ka? Robottorinri no senmonka ni kiku" [Is the use of unmanned drones a violation of the U.S. Constitution? We ask a robot ethics expert]. *Wall Street Journal*, July 26, 2012. http://jp.wsj.com/public/page/0_0_WJPP_7000-484015.html.

Kato, Hisakazu. 2019. "Can the Social Security System Be Sustained?" *Japan Times*, January 10, 2019. www.japantimes.co.jp/opinion/2019/01/10/commentary/japan-commentary/can-social-security-system-sustained.

Katsuno, Hirofumi. 2010. "Materializing Dreams: Humanity, Masculinity, and the Nation in Contemporary Japanese Robot Culture." PhD diss., University of Hawai'i.

Katsuno, Hirofumi. 2011. "The Robot's Heart: Tinkering with Humanity and Intimacy in Robot-Building." *Japanese Studies* 31 (1): 93–109.

Kavedžija, Iza. 2020. "Communities of Care and Zones of Abandonment in 'Super-aged' Japan." In *The Cultural Context of Aging: Worldwide Perspectives*, edited by Jay Sokolovsky, 211–30. Westport, CT: Praeger.

Kawaguchi, Daichi, and Mori Hiroaki. 2019. "The Labor Market in Japan, 2000–2018." RIETL (Japan Research Institute of Economy, Trade and Industry). www.rieti.go.jp/en/special/from-iza/016.html.

Khan, Aina J. 2020. "Could Robots in Care Homes Solve the UK's Social Care Crisis?" *Metro*, September 8, 2020. https://metro.co.uk/2020/09/08/could-robots-in-care-homes-solve-the-uks-social-care-crisis-13237787.

Kishi, Nobuhito. 2011. *Robotto ga nihon wo sukuu* [Robots will save Japan]. Tokyo: Bungeishunjū.

Kishimoto, Marimi. 2016. "How Are Robots Doing These Days?" *Nikkei Asian Review*, May 28, 2016. https://asia.nikkei.com/Tech-Science/Tech/How-are-robots-doing-these-days.

Kitano, Naho. 2006. "'Rinri': An Incitement towards the Existence of Robots in Japanese Society." *International Review of Information Ethics* 6:78–83. https://informationethics.ca/index.php/irie/article/view/143/141.

Kohlbacher, F., and B. Rabe. 2015. "Leading the Way into the Future: The Development of a (Lead) Market for Care Robotics in Japan." *International Journal of Technology, Policy and Management* 15 (1): 21–44.

Kubo, Akinori. 2015. *Robotto no jinruigaku: 20 seiki nihon no kikai to ningen* [Robot anthropology: 20th century Japanese machines and people]. Kyoto: Sekaishisōsha.

Lan, Pei-Chia. 2018. "Bridging Ethnic Differences for Cultural Intimacy: Production of Migrant Care Workers in Japan." *Critical Sociology* 44 (7–8): 1029–43. http://doi.org/10.1177/0896920517751591.

Lebra, Takie Sugiyama. 1976. *Japanese Patterns of Behavior*. Honolulu: University of Hawai'i Press.

Lechevalier, Sébastien, Yukio Ikeda, and Junichi Nishimura. 2010. "The Effect of Participation in Government Consortia on the R&D Productivity of Firms: A Case Study of Robot Technology in Japan." *Economics of Innovation and New Technology* 19 (8): 669–92. http://doi.org/10.1080/10438590902872903.

Leeson, Christina. 2017. "Anthropomorphic Robots on the Move: A Transformative Trajectory from Japan to Danish Healthcare." PhD diss., University of Copenhagen.

Lipp, Benjamin. 2019. "Interfacing RobotCare: On the Techno-politics of RobotCare." PhD diss., Technical University of Munich.

Liu, Sichun. 2017. "Humanoid Robot 'Pepper' Makes North American Debut at Cornell." *Cornell Daily Sun*, April 28, 2017. http://cornellsun.com/2017/04/28/humanoid-robot-pepper-makes-north-american-debut-at-cornell/.

Lopez, Mario. 2012. "Reconstituting the Affective Labour of Filipinos as Care Workers in Japan." *Global Networks* 12 (2): 252–68.

MacFarquhar, Larissa. 2018. "The Comforting Fictions of Dementia Care." *New Yorker*, October 1, 2018. www.newyorker.com/magazine/2018/10/08/the-comforting-fictions-of-dementia-care.

Maher, JaneMaree. 2009. "Accumulating Care: Mothers beyond the Conflicting Temporalities of Caring and Work." *Time & Society* 18 (2–3): 231–45.

Manyika, James, Susan Lund, Michael Chui, Jacques Bughin, Jonathan Woetzel, Parul Batra, Ryan Ko, and Saurabh Sanghvi. 2017. *Jobs Lost, Jobs Gained: Workforce Transitions in a Time of Automation*. New York: McKinsey Global Institute.

Marx, Patricia. 2018. "Learning to Love Robots." *New Yorker*, November 19, 2018. www.newyorker.com/magazine/2018/11/26/learning-to-love-robots.

McCurry, Justin. 2018. "System Error: Japan Cybersecurity Minister Admits He Has Never Used a Computer." *The Guardian*, November 15, 2018. www.theguardian.com/world/2018/nov/15/japan-cyber-security-ministernever-used-computer-yoshitaka-sakurada.

McLaughlin, Levi. 2013. "What Have Religious Groups Done after 3.11? Part 2: From Religious Mobilization to 'Spiritual Care.'" *Religion Compass* 7 (8): 309–25.

Meacham, Darian, and Matthew Studley. 2017. "Could a Robot Care? It's All in the Movement." In *Robot Ethics 2.0: From Autonomous Cars to Artificial Intelligence*, edited by Patrick Lin, Keith Abney, and Ryan Jenkins, 97–112. Oxford: Oxford University Press.

Meintjes, Ingrid. 2020. "Care in a Can." Paper presented at the Robophilosophy conference 2020.

Melkas, Helinä, Lea Hennala, Satu Pekkarinen, and Ville Kyrki. 2020. "Impacts of Robot Implementation on Care Personnel and Clients in Elderly-care Institutions." *International Journal of Medical Informatics*, 134, 104041. https://doi.org/10.1016/j.ijmedinf.2019.104041.

Ministry of Health, Labour and Welfare (MHLW). 2016. "Gyōmujōshippeihasseijōkyōtōchōsa" [Survey of the situation of the occurrence of occupational illnesses]. www.mhlw.go.jp/bunya/roudoukijun/anzeneisei11/h28.html.

Ministry of Health, Labour and Welfare (MHLW). 2017a. "Heisei 29 nendo kaigoh okenjigyōjōkyōhōkoku (nenpō)" [2017 Long-Term Care Insurance business status report (annual report)]. www.mhlw.go.jp/topics/kaigo/osirase/jigyo/17/index.html.

Ministry of Health, Labour and Welfare (MHLW). 2017b. "Kaigorōjinfukushishisetsu (sankōshiryō)" [Elderly care welfare institutions (reference materials)].
Ministry of Health, Labour and Welfare (MHLW). 2018. "Dai 7 ki kaigohokenjigyōkeikaku ni motozuku kaigojinzai no hitsuyōsū ni tsuite" [Concerning the required number of care staff based on the 7th Long-Term Care Insurance Plan]. www.mhlw.go.jp/stf/houdou/0000207323.html.
Ministry of Health, Labour and Welfare (MHLW). 2019. "Kaigosābisukibanseibi ni tsuite (sankōshiryō)" [Regarding care service infrastructure development (reference materials)]. www.mhlw.go.jp/content/12300000/000547178.pdf.
Ministry of Health, Labour and Welfare (MHLW). 2021. "Kaigokoyōkanrikaizen nado keikaku kaiseian ni tsuite" [Regarding the proposal to revise the plan for the improvement of care employment management and so on]. https://www.mhlw.go.jp/content/11601000/000744961.pdf.
Ministry of the Economy, Trade, and Industry (METI). 2019. "Miraiinobēshon WG kara no messēji" [Message from the future innovation working group]. www.meti.go.jp/press/2018/03/20190319006/20190319006-1.pdf.
Mol, Annemarie, Ingunn Moser, and Jeannette Pols. 2010. "Care: Putting Practice into Theory." In *Care in Practice: On Tinkering in Clinics, Homes and Farms*, edited by Annemarie Mol, Ingunn Moser, and Jeannette Pols, 7–25. New York: Columbia University Press.
Mori, Masahiro. 1970. "Bukimi no tani" [The uncanny valley]. *Enajī* [Energy] 7:33–35.
Mori, Masahiro. 1981. *The Buddha in the Robot*. Tokyo: Kosei Publishing.
Morris, Regan. 2015. "Why the 'Cute Robots' Don't Work for Rodney Brooks." BBC, November 17, 2015. www.bbc.com/news/business-34810552.
Nakada, Makoto. 2010. "Different Discussions on Roboethics and Information Ethics Based on Different Cultural Contexts (BA)." In *Cultural Attitudes towards Communication and Technology*, edited by F. Sudweeks, H. Hrachovec, and C. Ess, 300–15. Murdoch, Australia: Murdoch University.
Nakamura, Miri. 2007. "Marking Bodily Differences: Mechanized Bodies in Hirabayashi Hatsunosuke's 'Robot' and Early Showa Robot Literature." *Japan Forum* 19 (2): 169–90.
Nakayama, Shin. 2006. *Robotto ga nihon wo sukuu* [Robots will save Japan]. Tokyo: Tōyōkeizaishinpōsha.
Neven, Louis, and Alexander Peine. 2017. "From Triple Win to Triple Sin: How a Problematic Future Discourse Is Shaping the Way People Age with Technology." *Societies* 7 (3): 1–11.
Nickelsen, Niels Christian M. 2019. "Imagining and Tinkering with Assistive Robotics in Care for the Disabled." *Paladyn, Journal of Behavioral Robotics* 10 (1): 128–39. https://doi.org/10.1515/pjbr-2019-0009.
Nikkei Asia. 2014. "Next 10 Years Crucial for Japan's Nursing-Care Robot Industry." *Nikkei Asia*, November 5, 2014. https://asia.nikkei.com/Economy/Next-10-years-crucial-for-Japan-s-nursing-care-robot-industry.
Nippon.com. 2020. "Record 1.66 Million Foreign Workers in Japan in 2019 Society." Nippon.com, March 30, 2020. www.nippon.com/en/japan-data/h00676/record-1-66-million-foreign-workers-in-japan-in-2019.html.
Ochiai, Emiko. 2005. "The Ie (Family) in Global Perspective." In *A Companion to the Anthropology of Japan*, edited by Jennifer Robertson, 355–s79. Malden, MA: Blackwell.
O'Dea, Christopher. 2017. "Healthcare Japan: Longest-Term Investment." IPE Real Assets, January/February 2017. https://realassets.ipe.com/alternatives/healthcare-japan-longest-term-investment/10017355.article.

Ogawa, Reiko. 2021. "The Changing Face of Old-age Care in Japan." Paper presented at the Ludwig-Maximilians-Universität München Japan Center, December 13.
Okada, Michio. 2012. *Yowai robotto* [Weak robots]. Tokyo: Igakushoin.
Okada, Michio. 2016. "Human-Dependent Weak Robots for Creating Symbiotic Relations with Human." *Journal of the Robotics Society of Japan* 34:5, 299–303. http://doi.org/10.7210/jrsj.34.299.
Okamoto, Shinpei. 2013. "Nihon ni okeru robotto rinrigaku" [The study of robot ethics in Japan]. *Shakai to rinri* [Society and ethics] 28:5–20.
Ong, Sandy. 2020. "Will Robots and AI Take Our Jobs in COVID-19's Socially Distanced Era?" *New Scientist*, October 7, 2020.
Organisation for Economic Co-operation and Development (OECD). 2017. "Old-Age Dependency Ratio." In *Pensions at a Glance 2017: OECD and G20 Indicators*. Paris: OECD Publishing. http://doi.org/10.1787/pension_glance-2017-22-en.
Otsuki, Grant Jun. 2015. "Human and Machine in Formation: An Ethnographic Study of Communication and Humanness in a Wearable Technology Laboratory in Japan." PhD diss., University of Toronto.
Oudshoorn, Nelly. 2011. *Telecare Technologies and the Transformation of Healthcare*. London: Palgrave Macmillan.
Parks, Jennifer. 2010. "Lifting the Burden of Women's Care Work: Should Robots Replace the 'Human Touch'?" *Hypatia* 25:100–20. https://doi.org/10.1111/j.1527-2001.2009.01086.x
Peng, Ito. 2002. "Social Care in Crisis: Gender, Demography, and Welfare State Restructuring in Japan." *Social Politics: International Studies in Gender, State & Society* 9 (3): 411–43.
Peng, Ito. 2018. "From Care Work and Migration to Care Economy in East and Southeast Asian Contexts." Fondation France-Japon. April 4, 2018. http://ffj.ehess.fr/index/article/359/from-care-work-and-migration-to-care-economy-in-east-and-southeast-asian-contexts.html.
Peng, Ito, and Joseph Wong. 2010. "East Asia." In *The Oxford Handbook of the Welfare State*, edited by Francis G. Castles, Stephan Leibfried, Jane Lewis, Herbert Obinger, and Christopher Pierson, 656–70. Oxford: Oxford University Press.
Persson, Marcus, David Redmalm, and Clara Iversen. 2021. "Caregivers' Use of Robots and their Effect on Work Environment—A Scoping Review." *Journal of Technology in Human Services*. http://doi.org/10.1080/15228835.2021.2000554.
Plourde, Lorraine. 2014. "Cat Cafés, Affective Labor, and the Healing Boom in Japan." *Japanese Studies* 34 (2): 115–33.
Prochaska, James, and John Norcross. 2002. *Systems of Psychotherapy: A Transtheoretical Analysis*. 5th ed. Pacific Grove, CA: Brooks/Cole.
Read, Emily, Cora Woolsey, Chris A. McGibbon, and Colleen O'Connell. 2020. "Physiotherapists' Experiences Using the Ekso Bionic Exoskeleton with Patients in a Neurological Rehabilitation Hospital." *Rehabilitation Research and Practice* 2020: 2939573. https://doi.org/10.1155/2020/2939573.
Richardson, Kathleen. 2015. *An Anthropology of Robots and AI: Annihilation Anxiety and Machines*. London: Routledge.
Roberts, Glenda, and Hiroko Costantini. 2021. "The Work, Family and Care Nexus in Paris and Tokyo: Gender Equality and Well-Being among Urban Professionals." *Contemporary Japan*. http://doi.org/10.1080/18692729.2021.1925399.
Robertson, Jennifer. 2007. "Robo Sapiens Japanicus: Humanoid Robots and the Posthuman Family." *Critical Asian Studies* 39 (3): 369–98.
Robertson, Jennifer. 2010. "Gendering Humanoid Robots: Robo-Sexism in Japan." *Body & Society* 16 (2): 1–36.

Robertson, Jennifer. 2014. "Human Rights vs Robot Rights: Forecasts from Japan." *Critical Asian Studies* 46 (4): 571–98.
Robertson, Jennifer. 2018. *Robo Sapiens Japanicus: Robots, Gender, Family, and the Japanese Nation*. Oakland: University of California Press.
Roquet, Paul. 2009. "Ambient Literature and the Aesthetics of Calm: Mood Regulation in Contemporary Japanese Fiction." *Journal of Japanese Studies* 35 (1): 87–111.
Šabanović, Selma. 2014. "Inventing Japan's 'Robotics Culture': The Repeated Assembly of Science, Technology, and Culture in Social Robotics." *Social Studies of Science* 44 (3): 342–67.
Sado, Mitsuhiro, Akira Ninomiya, Ryo Shikimoto, Baku Ikeda, Toshiaki Baba, Kimio Yoshimura, et al. 2018. "The Estimated Cost of Dementia in Japan, the Most Aged Society in the World." *PLoS ONE* 13 (11): e0206508. http://doi.org/10.1371/journal.pone.0206508.
Sakanaka, Hidenori. 2007. "The Future of Japan's Immigration Policy: A Battle Diary." *Asia-Pacific Journal: Japan Focus* 5 (4): 1–9.
Santos, Julie, Jessica Yoon, and Davis Park. 2015. "PARO 6-Month Analysis: Front Porch Center for Innovation and Wellbeing." Paper presented at the 6th International Symposium on Robot Therapy with Seal Robot, PARO, AIST Tokyo Waterfront.
Savills. 2019. "Do the Golden Years Offer Golden Returns?" June 30, 2019. www.savills.com/prospects/sectors-apac-senior-housing.html.
Schodt, Frederik L. 1988. *Inside the Robot Kingdom: Japan, Mechatronics, and the Coming Robotopia*. Tokyo: Kodansha International.
Sharkey, Amanda, and Noel Sharkey. 2012. "Granny and the Robots: Ethical Issues in Robot Care for the Elderly." *Ethics and Information Technology* 14 (1): 27–40.
Sharkey, Noel, and Amanda Sharkey. 2012. "The Eldercare Factory." *Gerontology* 58 (3): 282–88.
Shibata, T., K. Wada, Y. Ikeda, and S. Šabanovic. 2008. "Tabulation and Analysis of Questionnaire Results of Subjective Evaluation of Seal Robot in Seven Countries." RO-MAN 2008—The 17th IEEE International Symposium on Robot and Human Interactive Communication, Munich, August 1–3, 2008.
Sone, Yūji. 2017. *Japanese Robot Culture: Performance, Imagination, and Modernity*. New York: Palgrave Macmillan.
Sparrow, Robert, and Linda Sparrow. 2006. "In the Hands of Machines? The Future of Aged Care." *Minds and Machines* 16 (2): 141–61.
Statistics Bureau of Japan. 2019. *Japan Statistical Yearbook 2019*. Tokyo: Statistics Bureau, Ministry of Internal Affairs and Communications. www.stat.go.jp/english/data/nenkan/index.html.
Sternsdorff-Cisterna, Nicolas. 2015. "Food after Fukushima: Risk and Scientific Citizenship in Japan." *American Anthropologist* 117 (3): 455–67.
Strauss, Delphine. 2016. "Robots Could Replace Migrant Workers, Says Think-Tank." *Financial Times*, July 4, 2016. www.ft.com/content/a1614f98-4123-11e6-9b66-0712b3873ae1?mhq5j=e7.
Świtek, Beata. 2014. "Representing the Alternative: Demographic Change, Migrant Eldercare Workers, and National Imagination in Japan." *Contemporary Japan* 26 (2): 263–80.
Świtek, Beata. 2016. *Reluctant Intimacies. Indonesian Eldercare Workers and National Imagination in Japan*. New York: Berghahn Books.
Tahhan, Diana Adis 2014. *The Japanese Family: Touch, Intimacy and Feeling*. London: Routledge.
Takahashi, Hara. 2016. "The Ghosts of Tsunami Dead and Kokoro No Kea in Japan's Religious Landscape." *Journal of Religion in Japan* 5 (2–3): 176–98.

Tanaka, Masayuki. 2018. "Peppākun sayonara. 8 warichō ga 'mō iranai'" [So long Pepper. More than 80% say "I don't need it anymore"]. *Dot Asahi*, October 25, 2018. https://dot.asahi.com/wa/2018102400011.html.
Thang, Leng Leng. 2001. *Generations in Touch: Linking the Old and Young in a Tokyo Neighborhood*. Ithaca, NY: Cornell University Press.
Thomas, Zoe. 2020. "Coronavirus: Will COVID-19 Speed Up the Use of Robots to Replace Human Workers?" BBC News, April 18, 2020. www.bbc.co.uk/news/technology-52340651.
Tronto, Joan. 2013. *Caring Democracy: Markets, Equality, and Justice*. New York: New York University Press.
Tsing, Anna Lowenhaupt. 2015. *The Mushroom at the End of the World: On the Possibility of Life in Capitalist Ruins*. Princeton, NJ: Princeton University Press.
Tsujimura, Mayuko, Hiroo Ide, Wenwei Yu, Naonori Kodate, et al. 2020. "The Essential Needs for Home-care Robots in Japan." *Journal of Enabling Technologies* 14 (4): 201–20. http://doi.org/10.1108/JET-03-2020-0008.
Turing, Alan M. 1950. "Computing Machinery and Intelligence." *Mind* 49:433–60.
Turkle, Sherry. 2011. *Alone Together*. New York: Basic Books.
Turkle, Sherry, Will Taggart, Cory D. Kidd, and Olivia Dasté. 2006. "Relational Artifacts with Children and Elders: The Complexities of Cybercompanionship." *Connection Science* 18 (4): 347–61.
Ueda, Hisatoshi, Shin-ichi Ito, Katsuya Sato, and Shoichiro Fujisawa. 2012. "Kaijosagyōchū no yōtsūchōsa to beddo kaijo futan hyōka" [Lower back pain survey during care work and bed care load assessment]. *Fukushi no machizukurikenkyū* [Journal of the Japanese Association for an Inclusive Society] 14 (2): A9-A17. https://www.jstage.jst.go.jp/article/jais/14/2/14_KJ00008157823/_pdf/-char/ja.
Ueno, Chizuko. 2011. *Ohitorisama no rōgo* [Alone in old age]. Tokyo: Bunshun Bunko.
UK Government. 2021. "Build Back Better: Our Plan for Health and Social Care." http://www.gov.uk/government/publications/build-back-better-our-plan-for-health-and-social-care.
University of Texas at Tyler Magazine. 2016. "A Personal Touch: UT Tyler Professor Explores Use of Robotic Pet in Treating Dementia." *University of Texas at Tyler Magazine*, Fall/Winter 2015, 16–19. https://issuu.com/hudsongraphics/docs/uttyler_fall winter/18.
U.S. Census Bureau. 2017. *National Population Projections Tables*. www.census.gov/data/tables/2017/demo/popproj/2017-summary-tables.html.
Vallès-Peris, Núria, and Miquel Domènech. 2020. "Roboticists' Imaginaries of Robots for Care: The Radical Imaginary as a Tool for an Ethical Discussion." *Engineering Studies* 12 (3): 157–76. http://doi.org/10.1080/19378629.2020.1821695.
Vallor, Shannon. 2015. "Moral Deskilling and Upskilling in a New Machine Age: Reflections on the Ambiguous Future of Character." *Philosophy & Technology* 28 (1): 107–24.
van Wynsberghe, Aimee. 2016. "Service Robots, Care Ethics, and Design." *Ethics and Information Technology* 18: 311–21. https://doi.org/10.1007/s10676-016-9409-x.
Vogt, Gabriele, and Anne-Sophie L. König. 2021. "Robotic Devices and ICT in Long-term Care in Japan: Their Potential and Limitations from a Workplace Perspective." *Contemporary Japan*. http://doi.org/10.1080/18692729.2021.2015846.
Vollset, Stein Emil, Emily Goren, Chun-Wei Yuan, et al. 2020. "Fertility, Mortality, Migration, and Population Scenarios for 195 Countries and Territories from 2017 to 2100: A Forecasting Analysis for the Global Burden of Disease Study." *The Lancet* 396, no. 10258: P1285–P1306. http://doi.org/10.1016/S0140-6736(20)30677-2.

Wagner, Cosima. 2013. *Robotopia Nipponica—Recherchen zur Akzeptanz von Robotern in Japan* [Robotopia nipponica—research on the acceptance of robots in Japan]. Marburg: Tectum.
Wajcman, Judy. 2015. *Pressed for Time: The Acceleration of Life in Digital Capitalism.* Chicago: University of Chicago Press.
Wajcman, Judy. 2016. "Pressed for Time: The Digital Transformation of Everyday Life: Huvudanförande vid Sociologidagarna i Uppsala 10–12 Mars 2016." *Sociologisk Forskning* 53 (2): 193–98. https://www.jstor.org/stable/24899037.
Weizenbaum, Joseph. 1976. *Computer Power and Human Reason: From Judgment to Calculation.* New York: Freeman.
White, Daniel. 2018. "Contemplating the Robotic Embrace: Introspection for Affective Anthropology." More-than-Human Worlds: A NatureCulture Blog Series, June 20. https://blognatureculture.wordpress.com/2018/06/18/contemplating-the-robotic-embrace/.
White, Daniel, and Hirofumi Katsuno. 2021. "Toward an Affective Sense of Life: Artificial Intelligence, Animacy, and Amusement at a Robot Pet Memorial Service in Japan. *Cultural Anthropology* 36 (2): 222–51.
Wilding, Raelene. 2018. "Resisting and Embracing Technologies of Aged Care: Representations and Practices of Older People." Paper presented at the Society for Social Studies of Science conference, Sydney, August 29–September 1.
Women's Budget Group (WBG). 2020. "Creating a Caring Economy: A Call to Action." https://wbg.org.uk/analysis/creating-a-caring-economy-a-call-to-action-2/.
Wright, James. 2019. "Robots vs Migrants? Reconfiguring the Future of Japanese Institutional Eldercare." *Critical Asian Studies* 51 (3): 331–54.
Wright, James. 2021. "Comparing Public Funding Approaches to the Development and Commercialization of Care Robots in the European Union and Japan." *Innovation: The European Journal of Social Science Research.* https://doi.org/10.1080/13511610.2021.1909460.
Yamasaki, Takashi. 2006. "Kango, kaigobunya ni okeru gaikokujin rōdōsha no ukeire mondai" [Problems regarding the acceptance of foreign workers in the nursing and care fields]. *Refarensu* [Reference] 2006 (2): 4–24.
Yamato, Reiko. 2006. "Changing Attitudes towards Elderly Dependence in Postwar Japan." *Current Sociology* 54, no. 2:273–91. http://doi.org/10.1177/0011392106056746.
Yano Research Institute. 2018. "Kaigorobotto shijō ni kansuru chōsa wo jisshi" [Survey conducted on the care robot market]. www.yano.co.jp/press-release/show/press_id/1960.
Yoshida, Reiji, and Sakura Murakami. 2018. "More than 345,000 Foreign Workers Predicted to Come to Japan under New Visas: Government." *Japan Times*, November 14, 2018. www.japantimes.co.jp/news/2018/11/14/national/politics-diplomacy/345000-foreign-workers-predicted-come-japan-new-visas-government.

Index

Page numbers in *italics* indicate illustrations.

Abe, Shinzō, 9–11, 30, 33, 37, 146, 153n3
active listening, 66, 67, 101, 102, 104–6, 157n8. *See also keichō* and *iyashi*
activities of daily life (ADL), 40, 76
Actroid F androids, 38, 47, 106
actuators, 7, 12, 56, 152n13
Agency for Medical Research and Development (AMED), Japan, 10, 11, 17, 40, 42, 154n15
AIBO, 40, 43, 96, 98, 99, 151n9, 157n7
AIST. *See* National Institute of Advanced Industrial Science and Technology
Aldebaran, 56, 122
algorithmic care, 19, 55–57, 138–39, 145
Allison, Anne, 154n2
Alphabet (company), 122, 159n2
Ambo, Phie, 110–11
animal therapy, 97, 100
Annual White Paper on the Aging Society (Cabinet Office), 22
anshin/anzen, 87, 88, 90, 93
Aotani Institution for the Elderly, 76
Aricept (donepezil), 100
Asian Anthropology, 17
ASIMO, 13, 15, 16, 152n16
Astro Boy (Tetsuwan atomu), 13, 14, 135, 152n14
Atelier Akihabara, 123, 124
Atlas robot, 47, 155n6
attitudes towards robots in Japan, 7–11, 13, 135–36, 152–53nn1–2
Aum Shinrikyō, 101

back pain resulting from lifting, 71, 80, 81, 83–85, 93, 156–57nn7–8, 156n5
Berthin, Michael, 9
Bethel, Diana, 72, 76
Boston Dynamics, 43, 47, 155nn6–7, 159n2
Breazeal, Cynthia, 96
Brooks, Rodney, 152n16, 155n12
Brynjolfsson, Erik, 7
Buch, Elana, 33

Cabinet Office, Japan, 22, 135
Campbell, John, 23

Čapek, Karel and Josef, 11, 134, 152n12
capitalism: commodification of care industry, 33; contradiction of, 5–6, 141, 144, 147; industrialized temporality under, 137–38; reproductive labor, need for, 4–7; social relations, commodification of, 103; time, space, and capacity issues at Sakura and, 78; valuation of care under, 147
care crisis and introduction of robots, 21–35; alternative approaches to sustainable aging, 144–47; commodification of care industry, 33; differing visions of eldercare automation at METI and MHLW, 34–35; discourse of, 21–23; family care patterns, changes in, 5, 6, 23–24, 26, 27, 28, 33; industrial robotics, deployment of, 24–25, 135, 153n6; introduction of care robots, 33–34; labor shortage in care market, 28–29, 32, 115, 118–20, 141; life expectancy, postwar increase in, 24; LTCI (Long-Term Care Insurance) system and, 10, 27–30, 33, 34, 35, 41; migrant care workers and, 6, 15, 20, 30–33, 154n12; welfare state, development of, 23–26, 154n10
care fragmentation, 129, 160n14
care homes: actual use of robots by, 134–35, 160n3; animal therapy, problems with, 100; eligibility for entry into, 29–30, 61–62; fixed incomes of, 30; government subsidies for care robots, 155–56n4; hospitals, older people living in, 25–26; as REITs (real estate investment trusts), 161n8; Robot Care Project/RIRC's interaction with, 39, 45–46, 58, 143. *See also* Sakura nursing care home
care workers: ambivalence toward foreign care workers, 119–20; dependence of robotic aids on, 128–32, 139–41; depictions of, 156n9; deskilling/upskilling and introduction of robots, 129–30, 140, 142; gender issues, 18, 28, 45, 60, 69, 71, 81, 153n8, 155n2; home care services for older adults, 29, 154n12; impact of robot use on, 20, 131–32; individualized/personalized care ethics of,

175

care workers (continued)
75–77, 91–94, 131, 136–39, 157n10; labor shortage, 28–29, 32, 115, 118–20, 141; lifting robots, resistance to, 82–83, 87–94, 140; low wages of, 28, 30, 60, 71, 127, 141, 155n3; Paro and, 92, 107–14, 128, 130, 140, 158n15; Pepper, responses to, 126–28; at Sakura, 59–61, 69–72, 156n7, 156n9; social status of, 28, 46, 81, 93, 141; technology, general acceptance of, 91–92; toughness/difficulty of job, 28, 68, 71, 77, 80. *See also* migrant care workers

CB2, 4
Chiba (Sakura care worker), 64, 67, 73, 91
China, 7, 14, 29, 30, 32, 51–52, 115, 146
Chūbachi, Ryōji, 37, 41
Coeckelbergh, Mark, 138
Comi Kuma, 98
commodification of care industry, 33
communication and social interaction: ethics of using robots with older adults for, 99–100; HRI (human-robot interaction), low status hierarchy of, 56–57, 98–99; individualized/personalized care at Sakura, 75–77; *iyashi* and *keichō*, discourses of, 101–7, 114, 157–58nn8–11; kinship relations between workers and residents at Sakura, 71–72, 74–75; perception as gendered and innate skill, 99; between residents and care workers at Sakura, 72–75; at RIRC, 43–46; between robotics researchers and care home workers/residents, 39, 45–46, 58; as therapeutic technique, 158n11; therapy, computerization of, 105–7; touch, in Japan, 73–74. *See also* Paro
Confucianism, 23, 27
Costantini, Hiroko, 153–54n9
cost of eldercare robots, 141–43
COVID-19 pandemic, 9, 21, 32, 74, 128, 146, 151n7, 156n4, 157n8
Critical Asian Studies, 17
"cultural odorlessness" of robots, 130
Cyberdyne, 13, 39, 82

Danely, Jason, 93
deep hanging out, 16–17
Deleuze, Gilles, 103
dementia: care robots and, 4; deception of dementia patients, ethics of, 113, 158n17; *keichō* and, 105; lack of interaction of researchers with persons with, 46; lifting older adult care patients, and use of Hug, 85–86, 91; Paro and, 98–102, 105, 108, 109–14, 158n11, 158n15; Pepper and, 124; prevalence of, 2; public awareness of challenges of, 24; at Sakura nursing care home, 58–59, 61, 62, 65–68, 70, 73; temporality for patients with, 136
Dentsu (advertising company), 153n2
Diego (Sakura care worker), 65, 91, 119, 156n5
DOCTOR, 106
Doi, Toshi, 157n7
Domenech, Miquel, 160n14
Donath, Judith, 99
Dumouchel, Paul, 100

economic partnership agreements (EPAs), 31, 32, 116, 118–20, 128
Eguchi, Dr., 44–45
eldercare and robotics in Japan, 1–20, 133–49; actual use of robots by care homes, 134–35, 160n3; alternative approaches to sustainable aging, 144–47; attitude towards robots and, 7–11, 13, 135–36, 152–53nn1–2; capitalism's need for reproductive labor and, 4–7; cost issues, 141–43; defining robots, 11–16; dependence on care workers/human labor, 128–32, 139–41; future of, 141–43, 147–49; government support for, 9–11; growing population percentage of older adults and, 1, 5, 22, 49; human care labor, impact of robot use on, 20, 131–32; international development of, 14; lived reality, distance of robotics from, 133–35; methodology of study, 16–20; as "necessary technology," 9, 22; reality versus vision of, 1, 11, 16, 128; techno-dystopia, U.S./European fears of, 7–8; time, space, and capacity issues, 77–79, 136–41. *See also* care crisis and introduction of robots; Hug; lifting older care patients; lifting robots; National Institute of Advanced Industrial Science and Technology; Paro; Pepper; Sakura nursing care home
ELIZA, 106
Engelberger, Joseph, 151n8
EngKey, 151n9
enka, 67, 156n6
entrepreneurship at AIST, 40–43
environmental catastrophe and population levels, 147, 161n5
ethics: of deception of dementia patients, 113, 158n17; individualized/personalized care ethics, 75–77, 91–94, 131, 136–39, 157n10; of *iyashi* and *keichō*, 102; of robotics, 47–50, 53, 155n8; of using communication robots with older adults, 99–100

INDEX   177

Europe and America: care crisis discourse in, 23, 147; defining robots in, 13; ethics of robotics in, 47, 155n6; Germany, domestic robotics industry in, 51; humanoid dummies based on standardized male American bodies, 54; increasing older adult population in, 2, 22; industrial robotics, impact of, 153n6; investments in eldercare robotics in, 19; lifting policies in, 81, 156n3; lifting robots in, 82; monitoring *(mimamori)* systems in, 40; Nordic welfare states as lucrative care markets, 50; Pepper in, 148; robot/migrant worker dichotomy in, 117; techno-dystopia/job loss, fears of, 7–8, 89
Everyday Robo Recreation, 125

family: informal/unpaid/family eldercare, 5, 6, 23–24, 27–28, 146, 153–54n9; kinship relations between care workers and residents, 71–72, 74–75, 93; skinship and familial touch, 73–74; traditional family structure and care patterns, 5, 6, 23–24, 26, 27, 28, 33, 146
Farson, Richard, 104
fax machine, Japan's continuing dependence on, 21, 91
fertility rates in Japan, 22, 24, 26, 161n5
Fifth-Generation Computer Systems, 41
Fifth Science and Technology Basic Plan, 50, 151n10
Ford, Martin, 7
foreign care workers. *See* migrant care workers
Forsythe, Diana, 161n6
Fraser, Nancy, 5
Frey, Carl Benedikt, 105
Frumer, Yulia, 152–53n1
fuan, 87
Fugaku supercomputer, 21
Fujimoto (Sakura care worker), 65
Fujita (Sakura care worker), 1, 66, 70, 71, 75–76, 108
Fujiwara (Sakura care worker), 71
Fukushima nuclear disaster, 88, 135, 155n12, 157n8
Fukuyama, Francis, 7

Gagné, Nana, 154n2
Galapagos effect, 51
garakē, 51
genba, 17, 20, 36
Genda, Yuji, 30
gender issues: care home residents, gender distribution of, 60; care homes, roboticists' interactions with, 45; communication and social interaction, perceived as gendered and innate skill, 99; fertility rates, decline in, 24; government dominated by older men, 153n4; humanoid dummies based on standardized male American bodies, 54; industrial robotics, impact of, 153n6; in paid care work, 18, 28, 45, 60, 69, 71, 81, 153n8, 155n2; robots, gender identities of, 122, 158n14; traditional family, changing views on, 23–24, 28; work, social interaction at, 154n2; workforce, women's entry into, 23–24, 28, 153n8
Gineste, Yves, 158n10
Gold Plan, 26, 27
Goto (Sakura care worker), 77
government subsidies for care robots, 155–56n4
Gundam robot, 14
Gygi, Fabio, 152–53n1

Hakuhodo (advertising company), 153n2
HAL, 3, 13, 39, 82
Haraway, Donna, 161n5
Hashimoto, Akiko, 26
Hasse, Cathrine, 158n15
Headquarters for Japan's Economic Revitalization, 10
hikikomi genshō, 139–40
hikikomori, 114
Hirukawa, Hirohisa, 38, 42, 44, 82, 98, 157n6
Holliday, Ian, 31
home care services for older adults, 29, 154n12
Honda, Miwako, 158n10
Honda, Yukio, 42
Hong Kong, 22, 36, 155n7
HRP-4C, 15
Hug, 3, 16, 20, 39, 82, 86–94, 87, 112, 128–31, 140. *See also* lifting older care patients; lifting robots
Human Augmentation Research Center, 154n1
Humanitude, 158n10
human-robot interaction (HRI), low status hierarchy of, 56–57, 98–99
human subjects, standardization of, 53–57

Ikeda (Sakura care worker), 66, 70
immigrant care workers. *See* migrant care workers
independence/individualism, in eldercare context, 22, 28, 46, 75, 151n4, 153n7
individualized/personalized care ethics, 75–77, 91–94, 131, 136–39, 157n10

INDEX

Indonesia, migrant care workers from, 31, 118
industrial robotics, 24–25, 135, 153n6
informal/unpaid/family eldercare, 5, 6, 23–24, 27–28, 146, 153–54n9
Innovation 25, 4, 33, 117
Inoue (Sakura resident), 65–66, 109
Institute of Population Aging, Oxford University, 98
Intelligent Systems research institute, AIST, 38, 42–43, 56, 82, 98, 154n1
International Classification of Functioning, Disability, and Health (ICF), 54, 55
International Organization for Standardization (ISO): ISO 13482 standard, 34, 50, 52, 53, 154n14; new robot services standard, 50–53
Ishiguro, Hiroshi, 13, 15, 153n2
Ishiguro, Nobu, 89, 92
Ishihara, Kohji, 48
Ishikawa (Sakura care worker), 71, 73, 78
Ito (Sakura resident), 110–14, *111*, 128, 140, 158n16
Iwabuchi, Koichi, 130
Iwasaki (Sakura care worker), 67, 69–70, 73, 80, 92, 114, 120
iyashi and *keichō*, 101–7, 114, 157–58nn8–11

Japan, robotics and eldercare in. *See* eldercare and robotics in Japan
Japanese Body Dimension Database, 54, 55
The Japanese Family Touch, Intimacy and Feeling (Tahhan), 73
Japanese Nursing Association, 31
Japan Industrial Robot Association, 24
Japan Machinery Federation, 9
Japan Nursing Federation, 31
Japan Robot Association (JARA), 9, 40, 42, 51
Jasanoff, Sheila, 152n15

Mr. K (manager at Sakura), 60–63, 71, 84, 90, 100, 109, 115, 118–20, 123, 127–31, 156–57nn8–9
Kabochan, 40, 43, 98, 102
kaigorobotto (care robot), 154n14
Katsuno, Hirofumi, 155n8
Kavedžija, Iza, 26
Kawasaki Aircraft/Heavy Industries, 24
keichō and *iyashi*, 101–7, 114, 157–58nn8–11
Kim, Sang-Hyun, 152n15
Kimura (Sakura resident), 66, 67, 108
Kirobo, 153n2
Kishi, Nobuhito, 151n6, 155n10
Kishida, Fumio, 146
Kismet, 96

Kitano, Naho, 48
Kobe earthquake, 101
Kohlbacher, Florian, 8–9
Kokokuma, 98
kokoro/kokoro no kea, 48, 101, 102, 107, 155n8
Kubo (Sakura care worker), 65, 126–27
Kubo, Akinori, 14
Kuma-tan, 98
Kurita, Yuki, 103

labor shortage in care market, 28–29, 32, 115, 118–20, 141
Leeson, Christina, 12, 109, 158n15
Liberal Democratic Party, Japan, 26, 136, 153n4
lifting older care patients: as aim of care robotics, 4, 81; back pain resulting from, 71, 80, 81, 83–85, 93, 156–57nn7–8, 156n5; as care worker activity, 59, 64, 69, 71, 74, 80–81; individualized/personalized care ethics and, 91–94, 157n10; patterns and practices of lifting at Sakura, 80, 83–86; safe patient handling policies, 81, 156n3; as touch, 85–86, 93, 130
lifting robots: care worker resistance to, 82–83, 87–94, 112; Hug, 3, 16, 20, 39, 82, 86–94, *87*, 112, 128–31, 140; introduction of Hug at Sakura, 86–91, *87*; older care patients' unease about, 86, 87, 89, 90–91; Robear, *15*; types of devices, 81–83
Lin, Patrick, 48
Lipp, Benjamin, 151n8
lived reality, distance of robotics from, 133–35; algorithmic care, concept of, 19, 55–57, 138–39, 145; care, RIRC engineers' versus Sakura care workers' views of, 79, 93, 99, 138–39, 145; care homes and actual older adults, RIRC's lack of interaction with, 39, 45–46, 58, 143; as research and development problem, 161n6; HRI (human-robot interaction), low status hierarchy of, 56–57, 98–99
Long-Term Care Insurance (LTCI) system, 10, 27–30, 33–35, 41, 75, 79, 97, 137, 146, 147
Lovot, 98, 99

Maeda (Sakura care worker), 70, 77, 108, 133, 160n1
Maher, JaneMaree, 137–38
manipulators, 12, 152n13
manzai, 67
Marescotti, Rosetta, 158n10

INDEX 179

Massachusetts Institute of Technology (MIT), 14, 42, 43, 95, 96, 98, 106, 107
Matsui (Sakura care worker), 76, 85, 88–89
Matsukoroid, 13, 153n2
Matsumoto, Yoshio, 12, 38, 40, 41, 43, 53–54, 81, 84, 90, 160n3
Matsuo (Sakura care worker), 67, 71, 111–12
McKinsey Global Institute, 7–8
Meacham, Darian, 100
Mechanical Love (film), 110–11
mental health crises, national, 101–2, 157n8
Mexico, migrant care workers from, 128
migrant care workers: ambivalence of Japanese care workers regarding, 119–20; care crisis in Japan and, 6, 15, 20, 30–33, 154n12; "cultural odorlessness" of robots versus, 130–31; EPAs (economic partnership agreements), 31, 32, 116, 118–20, 128; Pepper/robots and, 115–20, 130–31, 140, 141–42; rapid increase in, 146, 147; resistance to, in Japan, 116–20, 130; at Sakura, 65, 69, 115–20, 128, 156n5, 156n9; TITP (Technical Intern Training Program), 31–32, 128
military robotics, 47, 155n7
mimamori, 59, 127
Ministry of Economy, Trade, and Industry (METI), Japan: AIST and, 40, 41; care crisis in Japan and, 24, 30–31, 33, 34–35, 146, 154n10; care robotics, investment in, 9, 10, 11, 135, 141; EPAs (economic partnership agreements), 31, 32, 116, 118–20, 128; standardization, interest in, 50, 52
Ministry of Health, Labour, and Welfare (MHLW), Japan, 10, 21, 26, 30, 31, 34–35, 84
Misora, Hibari, 67
modernity, robots and nostalgia for, 7–11
Modern Times (Chaplin film), 133–34, 160n1
monitoring *(mimamori)* systems, 40
muen shakai, 28, 75, 101, 113, 130, 145
Murata (Sakura care worker), 119

nagare, 20, 63, 79, 83, 92, 136
Nakada, Makoto, 48
Nakamura (engineer at RIRC), 44, 45
Nakayama, Shin, 7
National Institute of Advanced Industrial Science and Technology (AIST), 36–57; author's fieldwork at, 16, 18, 19, 36; defined and described, 37; defining robots, 12; on differing visions of eldercare automation at METI and MHLW, 34–35; ethics of robotics at, 47–50, 53; foci of research and entrepreneurship at, 40–43; HRP-4C, 15; human subjects, standardized for testing purposes, 53–57; Intelligent Systems research institute, 38, 42–43, 56, 82, 98, 154n1; international development of specific robots and, 14, 96; lifting, focus on, 81; management style compared to Sakura care home, 61; organization of research groups at, 154n1; as research arm of METI, 10; silence and secrecy, culture of, 36, 39, 43–46, 53, 154n2; standardization of robots, interest in, 50–53; terms for care robots used at, 154n14; therapy, computerization of, 106–7; Tsukuba, as "science city," 36, 37–38; types of care robots studied at, 39–40. *See also* Robot Care Project; Robot Innovation Research Center
national mental health crises, 101–2, 157n8
National Robotics Strategy for Japan, 10
"necessary technology," eldercare robots viewed as, 9, 22, 48–49
Neven, Louis, 22, 46
New Energy and Industrial Technology Development Organization (NEDO), 8, 10, 11, 34
New Gold Plan, 27
New Robot Strategy, 81
Nomura (Sakura care worker), 70, 77, 109
novelty effect, 95, 98, 108–9, 143, 148, 157n6
nursing care homes. *See* care homes

Obama, Barack, 16
odor, 130
ohitorisama, 153n7
Okada, Michio, 113
Okamoto, Shinpei, 48
Onishi (Sakura care worker), 73
Osborne, Michael, 105
Otsuka (Sakura care worker), 65, 68, 70, 78, 84–85, 87, 89, 92, 119
Otsuki, Grant, 154n2
Oudshoorn, Nelly, 160n14
Oxford Martin School, 7

Palro, 45
Parks, Jennifer, 128–29
Paro, 16, 20, 95–114; animal therapy, as replacement for, 97, 100; automation of therapeutic practices and, 105–7; birth certificates and "citizenship" awarded to, 96, 157n1; care workers and, 92, 107–14, 128, 130, 140, 158n15; dementia patients and, 98–102, 105, 108, 109–14, 158n11,

Paro (*continued*)
158n15; description of, 95–98; entrainment entailed by, 139–40; ethics of using, 99–100; gender identity of, 158n14; HRI (human-robot interaction), low status hierarchy of, 98–99; Hug compared, 95, 96, 98, 109, 112; images of, *96, 111*; introduction at Sakura, *96*, 100–101, 107–10, *111*; *iyashi* and *keichō*, discourses of, 101–7, 114, 157–58nn8–11; manufacture and development of, 14; older adult care patients' response to, *96*, 108–12, *111*, 140; Pepper compared, 95, 96, 98, 109; at Robot Care Project, RIRC, and AIST, 40, 43; as "social robot," 43, 87, 98–100; as therapeutic robot, 95–98, 157n3
Peine, Alexander, 22, 46
Peng, Ito, 6, 25, 154n10
Pepper, 3, 16, 20, 115–32; advertising companies, robot promotion by, 153n2; autonomy, myth of, 123–24; care workers' responses to, 128–32; dependence on care workers, 126–32, 140; described, 120–24; in Europe and America, 148; Haikai mimamori app, 151n1; Hug compared, 131; images of, *87, 121, 126*; labor shortage in Japan and, 118–20; manufacture and development of, 14; migrant care workers and, 115–20, 130–31, 140; multipurpose humanoid, designed as, 122, 123, 151n9, 159nn6–7; obsolescence of, 147–48; officiating at robot pet memorial services, 21; older adult residents' responses to, 125–26, 131; Paro compared, 95, 96, 98, 109; recreational activities and, 115–16, 123, 125–27, *126*, 130, 131; at Robot Care Project and RIRC, 38, 40; Romeo compared, 56; at Sakura, 1, 63, *87*, 109, 120, 123, 124–28, *126*; as "social robot," 43, 45, 87, 120, 122–23; technical difficulties with, 1, 124–25, 127, 159n8; in use, versus promotional video, 1; Wandering Monitoring app, 1, 159n8
Pepper World, 124
personalized/individualized care ethics, 75–77, 91–94, 131, 136–39, 157n10
Philippines, migrant care workers from, 31, 69, 115, 118, 128
Plourde, Lorraine, 103, 114
popular culture, robots in, 13, 135, 152n14
population: environmental catastrophe and, 147, 161n5; growing population percentage of older adults, 1, 2, 5, 22, 49
post-traumatic stress disorder (PTSD), 106–7, 158nn11–12

Project for the Development and Standardization of Robot Care Devices, 11
Project for the Practical Utilization of Personal Care Robots, 33–34

Rabe, Benjamin, 9
reactionary postmodernism, 152n11
real estate investment trusts (REITs), 161n8
recreational activities and robots, 115–16, 123, 125–27, *126*, 130, 131
reproductive labor, capitalism's need for, 4–7
Resyone, 39
retro-tech, 10
Richardson, Kathleen, 99
Rise of the Robots (Ford), 7
Ritsuko's Rexercise, 125, 159n9
Robear, *15*
Roberts, Glenda, 153–54n9
Robertson, Jennifer, 10, 54, 116, 131, 152n11, 155n9, 157n1
Robohelper Sasuke, 82
Robo Sapiens Japanicus (Robertson), 10
robot, etymology of, 11, 134, 152n12
Robot Anthropology (Akinori), 14
Robot Care Project, 11; AMED and, 42, 154n15; author's research work on, 16, 19, 36; care homes, interaction with, 39, 45; crisis of care in Japan and, 34, 35; ethics of, 49; future of care robotics and, 142–43; HRI (human-robot interaction), low status hierarchy of, 56–57; lifting as focus of, 81–82; METI and MHLW, differing views of robots at, 35; objectives at, 39–40, 49; preceding related projects, 34; silence and secrecy, AIST culture of, 39, 44; technology companies' wariness of, 161n4; as umbrella project, 43
Robot Ethics (Lin, Abney, and Jenkins, eds.), 48
robotics and eldercare. *See* eldercare and robotics in Japan
Robot Innovation Research Center (RIRC), 12, 38; care, view of, 79, 93, 99, 138–39, 145; communication and social interaction at, 43–46; ethics of robotics at, 47–50; future of care robotics and, 143; HRI (human-robot interaction), low status hierarchy of, 56–57; human informatics at, 54; lived reality of care homes, lack of connection to, 39, 45–46, 58; objectives of Robot Care Project, implementation of, 39–40; organization of research groups at AIST and, 154n1; research entrepreneurship at, 41
Robot Revolution Initiative, 9–10

Robot Revolution Realization Council, 37, 155n9
Robot Safety Testing Center, 34, 40, 52, 53–57
"robot season," 38, 53
Robots Will Save Japan (Kishi), 155n10
Robots Will Save Japan (Nakayama), 7
Robot Town Sagami, 117
Rogers, Carl, 104
Romeo, 56
Roomba, 13, 62, 155n12
Roquet, Paul, 102, 103, 114
RT2 walker, 39
Rubin, Andy, 159n2
R.U.R. (*Rossum's Universal Robots*; Čapek), 11, 134, 135

Šabanović, Selma, 95–96
Saito (Sakura resident), 69
Sakanaka, Hidenori, 117
Sakura nursing care home, 58–79; author's research methodology and, 16, 18; care workers at, 59–61, 69–72, 156n7, 156n9; communication with residents at, 72–75; day care center at, 155n1; description of, 58–60; Hug, introduction of, 86–91, *87*, 109; individualized/personalized care ethics at, 75–77, 91–94, 136–39; joking and teasing at, 65–66, 74, 80, 113; *keichō* at, 67, 104–5; kinship relations between workers and residents, 71–72, 74–75, 93; labor shortage at, 118–20; lifting patterns and practices at, 80, 83–86; management at, 60–61; migrant care workers at, 65, 69, 115–20, 128, 156n5, 156n9; *Modern Times* (Chaplin film) at, 133–34; *nagare* or flow of daily life at, 59–60, 63–69, 83, 136–37; Paro, introduction of, *96*, 100–101, 107–10, *111*; Pepper at, 1, 63, *87*, 109, 120, 123, 124–28, *126*; residents and admission process at, 59, 61–62; robots, introduction of, 60–63, 68, 79, 100–101; staff and management at, 59–61; technological regulation of life at, 91; time, space, and capacity issues, 77–79, 136–41; touch and tactile contact at, 73–74, 85–86, 93, 130
Schodt, Frederik, 25
The Second Machine Age (Brynjolfsson), 7
secrecy and silence, culture of, at AIST, 36, 39, 43–46, 53, 154n2
seikatsushienrobotto (lifestyle support robot), 154n14
Sharkey, Amanda, 6
Sharkey, Noel, 6
Shibata, Takanori, 42–43, 157n3, 95–97, 99

shiritori, 67
silence and secrecy, culture of, at AIST, 36, 39, 43–46, 53, 154n2
Silver Columbia Plan, 31
SimSensei, 106, 158n11
Sixth Science and Technology Basic Plan, 151n10
skinship, 73–74
smell, 130
Society 5.0, 10, 151n10
sociotechnical imaginaries, 14, 145, 152n15
SoftBank Robotics, 40, 56, 98, 122–27, 129, 148, 155n7, 159n3, 159n8
Son, Masayoshi, 122
Sone, Yuji, 96, 109, 113, 139–40
South Korea, 7, 22, 51, 118, 151n9
space, time, and capacity issues, 77–79, 136–41
Special Project to Support the Introduction of Care Robots, 34
standardization: of human subjects, for testing purposes, 53–57; ISO 13482 standard, 34, 154n14; Japanese (JIS) standard Y1001, 50; new robot services standard, 50–53; TC299 (technical committee for personal care robots), 50–51
Sternsdorff-Cisterna, Nicholas, 88
Studley, Matthew, 100
subsidies for care robots, 155–56n4
Suga, Yoshihide, 146, 153n4
Suzuki (Sakura resident), 1
Świtek, Beata, 32

Tahhan, Diana, 73
Takahashi (Sakura resident), 66, 110
Takeshita, Noboru, 26
Taylor (RIRC engineer), 51–52
TC299 (technical committee for personal care robots), 50–51
Technical Intern Training Program (TITP), 31–32, 128
techno-animism, 152–53n1
techno-orientalism, 15, 17, 21, 22, 148
Teddy Ears volunteer group, 104–5
Telenoid, 12, 109, 158n13
Thang, Leng Leng, 72
therapy: animal therapy, Paro as replacement for, 97, 100; automation of, 105–7; communication therapy techniques and technologies, 158n11; *iyashi* and *keichō*, discourses of, 101–7, 114, 157–58nn8–11; national mental health crises in Japan and, 101–2, 157n8; Paro, as therapeutic robot, 95–98

time, space, and capacity issues, 77–79, 136–41
touch and tactile contact, 73–74, 85–86, 93, 130, 136
Traphagan, John, 26
Tronto, Joan, 5, 28
Tsing, Anna, 5, 130
Tsukuba, 36, 37–38
Turing, Alan, and Turing test, 152n17
Turkle, Sherry, 99, 100, 106, 157n5

Ueno, Chizuko, 28, 153n7
uncanny valley effect, 14
United States. *See* Europe and America
unpaid/informal/family eldercare, 5, 6, 23–24, 27–28, 146, 153–54n9
U.S. National Bureau of Economic Research, 132

Valles-Peris, Núria, 160n14
Vallor, Shannon, 146
van Wynsberghe, Aimee, 161n7
Vietnam, migrant care workers from, 31, 32, 69, 118

Wabot House, 54
Wagner, Cosima, 153n6

Wajcman, Judy, 160n13
Wandering Monitoring app (Pepper), 1, 159n8
Watanabe (RIRC researcher), 57
Weizenbaum, Joseph, 106
welfare equipment (*fukushiyōgu*), care robots classified as, 34
welfare state, development of, 23–26, 154n10
Wilding, Raelene, 46
Wizard of Oz technique, 16, 152n17
women. *See* gender issues
Women's Budget Group, 146
Wong, Joseph, 154n10
work-life balance, concept of, 153–54n9
World Health Organization (WHO), 54

Yamaguchi (RIRC team member), 45
Yamamoto (RIRC engineer), 48–49
Yamamoto (Sakura resident), 80
Yamato, Reiko, 27
Yamazaki, Ritsuko, 159n9
yarushikanai, 48
Yoshida, Kenichi, 159n3
yoyū, 78, 93, 94, 102, 128, 136, 140, 148–49
Yurugi, Yoshiko, 8
yutori, 77, 78, 92

CPSIA information can be obtained
at www.ICGtesting.com
Printed in the USA
LVHW042322170223
739827LV00026B/1324/J